T0281425

# Band-Notch Characteristics in Ultra-Wideband Antennas

# Band-Notch Characteristics in Ultra-Wideband Antennas

Taimoor Khan and Yahia M. M. Antar

CRC Press
Taylor & Francis Group
Boca Raton London New York

CRC Press is an imprint of the
Taylor & Francis Group, an **informa** business

First edition published 2021
by CRC Press
6000 Broken Sound Parkway NW, Suite 300, Boca Raton, FL 33487-2742

and by CRC Press
2 Park Square, Milton Park, Abingdon, Oxon, OX14 4RN

---

### Library of Congress Cataloging-in-Publication Data

---

Names: Khan, Taimoor, author. | Antar, Yahia, author.
Title: Band-notch characteristics in ultra-wideband antennas / Taimoor Khan and Yahia M.M. Antar.
Description: First edition. | Boca Raton, FL : CRC Press, 2021. | Includes bibliographical references and index.
Identifiers: LCCN 2020052330 (print) | LCCN 2020052331 (ebook) | ISBN 9780367754723 (hardback) | ISBN 9781003163008 (ebook)
Subjects: LCSH: Ultra-wideband antennas. | Microstrip antennas--Design and construction.
Classification: LCC TK7871.67.U45 K43 2021 (print) | LCC TK7871.67.U45 (ebook) | DDC 621.3841/35--dc23
LC record available at https://lccn.loc.gov/2020052330
LC ebook record available at https://lccn.loc.gov/2020052331

---

ISBN: 978-0-367-75472-3 (hbk)
ISBN: 978-0-367-75569-0 (pbk)
ISBN: 978-1-003-16300-8 (ebk)

Typeset in Palatino
by KnowledgeWorks Global Ltd.

# Contents

# List of Figures

# List of Tables

# *Preface*

Advancement in wireless technology has altered our lives in past few decades. The technological enhancement has allowed us to avail new services for mobile communication such as voice, audio, video, and data services. Further, it has also helped us in accomplishing a faster data rate between portable devices and computers within a short range. This high-speed data rate can be enlarged by increasing the transmission power or employment of wide bandwidth. However, many portable devices operating with wireless technology are battery-powered, thus a large frequency bandwidth will be the solution for achieving high data rate. In this perspective, ultra-wideband (UWB) technology is a revolutionary approach in the field of wireless communication due to its high-speed data rate and excellent immunity to multi-path interference. The UWB research field has received great amount of interest since the decision by the Federal Communications Commission (FCC) in February 2002 authorizing the emission of very low-power spectral density in a bandwidth of 7.5 GHz going from 3.1 GHz to 10.6 GHz. Since then, UWB technology has been considered as one of the most promising technologies used in various applications such as radar, sensing, and military communications. In the development of UWB system, an antenna plays a significant role. Thus, a practical UWB antenna should be designed with better impedance matching, compact size, low cost, omni-directional radiation pattern, and flat group delay over a UWB region. Among different types of antennas, printed antennas are usually preferred for UWB technology-based communication systems because of their compact size, low profile, and easy integration facilities. These antennas help to make the system more robust and also reduce the implementation cost. Therefore, much effort has been devoted to the design of printed UWB antennas.

However, an underlying challenge in UWB antenna is to avoid interference due to some existing technologies that share the frequency bands within the UWB regulation standards. These are WiMAX (3.3–3.7 GHz band), WLAN (5.15–5.35 GHz and 5.725–5.825 GHz bands), X-band satellite downlink (7.25–7.75 GHz band), X-band ITU (8.025–8.4 GHz band), etc. Thus, it is essential to reject the interference with the existing wireless technologies.

The rapid growth of UWB technology over few decades has gained a lot of attention among researcher community. Traditionally, it has been used for radar imaging, but recently due to technological advancements, it allows users to access high-data communication services in cost of low-power utilizations. As an essential part of communication systems, antenna has attained deep attention among researchers and academicians as well. Among them, printed monopole antennas are very much popular because of their portable size, low weight, and ease of fabrication. However, antenna design for UWB applications should be capable of avoiding interferences with other existing narrow-band radio communication systems. Moreover, UWB antenna also suffers from multi-path fading, which is conquered by adopting the concept of multi-input multi-output (MIMO) technology. In this book, a comprehensive review has been carried out on UWB and UWB MIMO antennas with band-notched characteristics. This is first of its kind book on the emerging topic to the best of authors' knowledge.

The overall organization of the book is broadly classified into seven chapters as follows: Chapter 1 describes the overall introduction, followed by Single Band-Notched UWB Antennas (Chapter 2), Dual Band-notched UWB Antennas (Chapter 3), Multi Band-Notched UWB Antennas (Chapter 4), Band-Notched UWB MIMO Antennas (Chapter 5), Reconfigurable Band-Notched UWB Antennas (Chapter 6), and Tunable Band-Notched UWB Antennas (Chapter 7), respectively. Finally, some advanced applications of Ultrawideband Systems are also described in Chapter 8.

**Taimoor Khan and Yahia M. M. Antar**

# Acknowledgement

Dr. Taimoor Khan declares that the work embodied in this book was supported by the Science and Engineering Research Board (SERB) Govt. of India (Research Grant No. **SB/S3/EECE/093/2016 Dated −27/12/2016**).

# 1

## Band-Notched UWB Antennas

### 1.1 Introduction

Advancement in wireless technology has altered our lives in the past few decades. The technological enhancements have allowed us to avail new services for mobile communication such as voice, audio, video, and data services. Further, it has also helped us in accomplishing a faster data rate between portable devices and computers within a short range. This high-speed data rate can be enlarged by increasing the transmission power or employment of wide bandwidth. However, many portable devices operating with wireless technology are battery-powered, thus a large frequency bandwidth will be the solution for achieving high data rate [1]. In this perspective, ultra-wideband (UWB) technology is a revolutionary approach in the field of wireless communication due to its high-speed data rate and excellent immunity to multi-path interference. The UWB research field has received great amount of interest since the decision by the Federal Communications Commission (FCC) in February 2002 authorizing the emission of very low-power spectral density in a bandwidth of 7.5 GHz going from 3.1 GHz to 10.6 GHz [2, 3]. Since then, UWB technology has been considered as one of the most promising technologies used in various applications such as radar, sensing, and military communications [4].

In the development of UWB communication system, an antenna plays a significant role. Thus, a practical UWB antenna should be designed with better impedance matching, compact size, low cost, omni-directional radiation pattern, and flat group delay over the UWB region [5]. Among different type of antennas, printed antennas are usually preferred for UWB technology-based communication systems because of their compact size, low profile, and easy integration facilities. These antennas help to make the system more robust and also reduce the implementation cost. Therefore, much effort has been devoted to the design of printed UWB antennas [6]. However, an underlying challenge in UWB antenna is to avoid interference due to some existing technologies that share the frequency bands within the UWB regulation standards. These are Worldwide Interoperability for Microwave Access (WiMAX) (3.3–3.7 GHz band), Wireless Local Area Network (WLAN) (5.15–5.35 GHz and 5.725–5.825 GHz bands), C-band satellite downlink (7.25–7.75 GHz band), X-band (8.025–8.4 GHz band), etc. Thus, it is essential to reject the interference with the existing wireless technologies.

In order to investigate different techniques developed for introducing notched behavior in UWB antennas, a review has been carried out in this chapter based on the publications available in literature [7–100]. The authors have tried to bring the developments that have taken place related to designing of band-notched antennas in past few years and also to acknowledge the novel contribution of the researchers. The organization of this chapter has been done as follows: Section 1.1 describes the concept of band-notch in UWB antennas. Section 1.2 analyzes UWB antennas with single notch [7–37] characteristics followed by analysis of dual [38–57] and multi- [58–74] notch antennas in Section 1.3 and Section 1.4, respectively. The discussion also includes description of different reconfigurable and tunable configurations with notched bands in Section 1.5 [75–90]. Further, Section 1.6 [91–100] describes UWB band-notch multi-input multi-output (MIMO) antennas. A summary of this review process is highlighted in Section 1.7. Finally, Section 1.8 presents conclusion for this chapter, followed by the list of references.

## 1.2 Concept of Band-Notched Filtering in UWB Antennas

As mentioned in Section 1.1, UWB systems cover a wide frequency range, and thus, they interfere with the existing narrow-band communication systems. To address the interference problem in UWB communication, it is necessary to filter out the overlapping bands. The traditional method that is followed to suppress such potential interferences is to connect the antenna with narrow-band band-stop filters [101]. Figure 1.1 describes such a frequency notched antenna system.

The spectral response of band-stop filter reveals that it allows UWB frequency range (Figure 1.2(a)) to pass except at two frequencies ($f_1$ and $f_2$). Thus, these two frequencies are notched out in the spectral response of UWB system, as shown in Figure 1.2(b). However, such an approach fails to achieve

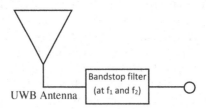

**FIGURE 1.1**
Notched ultra-wideband antenna system using bandstop filter.

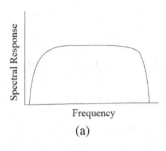
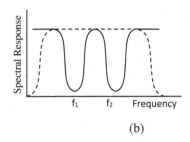

**FIGURE 1.2**
Development of frequency notched system using bandstop filters in terms of spectral response (a) ultra-wideband (UWB) antenna and (b) notched UWB system.

compactness and, at the same time, increases complexity and cost of the radio-frequency (RF)-front-end systems [102]. Thus, UWB antennas with band-rejection functions inherited in it are preferred. Such types of antennas are developed by introducing narrow-band resonant structures in them [103].

The addition of resonant structures results in band-notched characteristics in two ways. First, the impedance matching gets degraded by installing the resonant configuration, resulting in less energy transmission into/reception from the free space. Second, the current distribution on the antenna gets changed as a result of the addition of resonant structure, which may cause the canceling of radiation in the far-field zone [104]. Various techniques, thus, have been proposed to realize such resonant structures for notch characteristics which have been comprehensively analyzed in subsequent sections.

## 1.3 Single Band-Notched Characteristics

UWB antennas with single band-rejection characteristics are designed to avoid interference from the existing different narrow-band radio technologies operating at frequencies covered by the UWB band or its nearby frequencies. Different methods such as using slotted radiating or ground surface [7–23], placing parasitic patches [24–27] or installing open-ended stubs to the radiating patch [28–31], using electromagnetic bandgap (EBG) structure [32, 33], and multiple other techniques [34–37] have been proposed for obtaining single band-notched characteristics. One of the common techniques for developing notch performance is to embed slotted geometry in the antenna structure which changes the current flow, thereby resulting in a notch at a particular frequency. Some V-shaped slotted radiating surface-based monopole antenna geometries such as those discussed in [7, 8] and [9] have been proposed for generating single band-notched characteristics. Again, U-shaped slotted radiating surface-based monopole antennas, as given in

[10-14] are suggested for rejecting different bands, respectively. In [10], a second configuration of notched antenna is proposed with C-shaped slot. There are some other C-shaped slotted radiating surface-based monopole antennas as discussed in [15, 16]. In [17], a second configuration has also been proposed by implementing half wavelength slot on the ground plane. Rakluea et al. [18] have etched a rectangular slot in the ground plane for rejecting the undesired frequency band. An antenna configuration with meandered grounded stub is designed in [19], whereas shovel-shaped defected ground structure (DGS) structure is proposed in [20] for rejecting the undesirable WLAN band. A fractal UWB antenna using third iterative modified Sierpinski carpet pattern slots with U-shaped slots etched on the ground plane for rejecting 5–6 GHz WLAN band have been proposed by Biswas et al. [21]. Then ground plane of a Vivaldi antenna is introduced with a capacitively loaded loop (CLL) resonator to effectively notch the 5–5.9 GHz frequency band in [22]. Two different configurations of single band-notched antenna are obtained by means of C-shaped face-to-face and C-shaped back-to-back configurations [23].

Moreover, one of the popular approaches to develop notch property is to incorporate parasitic elements of different geometries either in the radiating plane or at the bottom plane of the antenna [24-27]. Sobli et al. [24] have used a slotted parasitic patch introduced at the bottom layer of the antenna for obtaining notch functioning. An H-shaped slot is etched from the parasitic patch placed at the backside of UWB antenna to have a single notch [25]. Inverted cup-shaped parasitic strip is added to the slotted ground in [26] to accomplish notched band behavior, whereas semi-octagonal parasitic strip is added beside the radiating patch in [27] to eliminate the interference with IEEE 802.11a services. Hong et al. [28] have designed a simple and compact microstrip-fed UWB antenna by placing a small rectangular strip bar at the center of two monopoles for obtaining band-notch characteristics. In [29], slotted ground plane is loaded with quarter-wavelength open-ended stubs to achieve band-notched behavior. Ghosh [30] has proposed a planar modified circular ring antenna loaded with a rectangular tuning stub for band-notch performance. Similarly, in order to achieve the band-rejection characteristic, Li et al. [31] have implemented a tuning stub in the middle of a fork-like patch. Further, the EBG structures are also used in UWB antennas to reject the interference. Yazdi et al. [32] have designed a new compact UWB circular monopole antenna with band rejection at 5.5 GHz. This rejection band is created by means of a mushroom-type EBG structure. An equivalent circuit model is also employed to investigate the stopband characteristic of the EBG. Similarly, Jaglan et al. [33] have proposed a circular monopole antenna for UWB applications with band-notch property. Antenna utilizes modified mushroom-type EBG structures to achieve band-notch property at WLAN 5–6 GHz band. Besides these, some other techniques [34–37] are also proposed for obtaining single notch characteristics in case of UWB antennas. A UWB dipole antenna with flat and sharp band-notch characteristics is proposed for obtaining both WLAN band notch and a compact UWB RF-front

**TABLE 1.1**

Techniques Used for Obtaining Single Band-notched Characteristics

| Technique | | Nature of Element with Cited Reference |
|---|---|---|
| Slots | On radiating surface | V-shaped slot [7–9], U-shaped slot [10, 14], C-shaped slot [10], [15, 16] |
| | In grounding surface | Open ended slot [17], rectangular slot [18], meandered ground slot [19], shovel-shaped DGS [20], U-shaped slot [21], capacitively loaded loop [22], C-shaped slot [23] |
| Parasitic-type elements | — | Slotted parasitic patch [24, 25], inverted-cup shape strip [26], semi-octagonal strip [27], |
| Stub-type elements | — | Rectangular stub [28–31] |
| Electromagnetic bandgap structures | | Mushroom [32], modified mushroom [33] |

structure [34]. A new configuration of monopole antenna has been designed in [35] where the band-notch characteristic is obtained by placing small trapezoidal antenna near the original planar trapezoidal antenna. Chu et al. [36] have proposed two wideband antennas with a shared aperture with band-notch characteristics in 820–1200 MHz band. Ghobadi et al. [37] have proposed two shorted quarter wavelength L-shaped stubs connected to ground plane via holes for generating frequency notch performance.

Thus, based on the available literature [7–37], different techniques implemented for obtaining single notch are summarized in Table 1.1. Further, the literature [7–37] is also characterized based on the feed mechanisms: Microstrip and co-planar waveguide (CPW) feeding used on different radiating element configurations and this is illustrated in Figure 1.3.

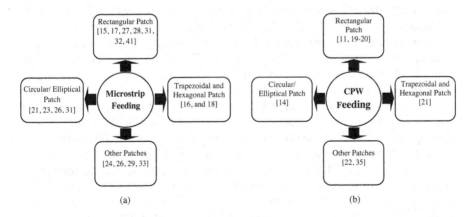

**FIGURE 1.3**

Categorization of antenna configurations used for single notch characteristics based on (a) microstrip feeding and (b) CPW feeding.

## 1.4 Dual Band-Notched Characteristics

At present, UWB antennas with dual band-notched characteristics are studied extensively to avoid interferences with the existing narrow-band communication systems. Usually, a single parasitic element or slot can generate only one notched band and fails to meet the requirements of mitigating multiple interferences. In order to realize a dual band-notched UWB antenna, multiple structures of slotted geometries [38–46], parasitic elements [47–51], EBG structure [52, 53], and some others [54–57] are commonly used. By etching slots of different shapes on the radiating patches [38–46], various configurations of UWB antennas are proposed. A pair of C-shaped slotted geometries on the radiating patch is used in [38] and [39] for obtaining band-notch characteristics. Yu et al. have etched L-shaped slots from the radiator patches where the antenna is configured based on two crossed tapered substrates [40]. Dual U-shaped slotted geometries are embedded in [41] to achieve dual notch properties. Mehranpour et al. [42] and Yadav et al. [43] have introduced dual notched-band antenna based on the L-shaped slots along with E and U slots, respectively. Jalil et al. [44] have used a couple of altered H-shaped slots for creating dual band-notch. Zhao et al. [45] have used arc-shaped slots implanted on a disk-shaped patch for obtaining dual notch characteristics. Avistika et al. [46] have employed two open-ended quarter wavelength slots on the radiating patch for rejecting WiMAX and X-band.

Introduction of parasitic elements can also contribute toward establishing dual notch attributes [47–51]. U-shaped slot with an I-shaped parasitic strip is employed in [47], and similarly, U-shaped slot and a pair of C-shaped parasitic stubs for generating dual band rejection are implemented in [48] and [49]. In [50], a single C-shaped slot is used to obtain first notch, whereas second notch is obtained by inserting T-shaped parasitic structure. Two coupled parasitic stepped impedance resonators (SIRs) are employed by Guang et al. [51] for rejecting dual frequency band. Recently, EBG structures are being used to reject the interference from the specified UWB frequency range [52, 53]. A novel design technique has been introduced by Wang et al. [52] which combines mushroom-shaped EBG structures and a slot together for producing dual notched-band characteristics at WLAN (4.8–5.9 GHz), and X-band downlink satellite communication band (7.1–7.8 GHz). It is also observed that on implementation of slot, the antenna not only creates its own notched band but also enhances the filtering performance at the other notched band generated by EBG structures. A good wideband radiation performance from 2.64 GHz to 12.9 GHz has been achieved. A new slitted EBG structure has been designed and implemented to filter out interfering frequencies within UWB frequency range. The EBG structure is placed near the feed line of circular patch UWB antenna that leads to generation of notches for applications in WiMAX and WLAN bands [53].

Some other techniques [54–57] are also used for establishing dual notch properties. Qi-kun et al. [54] have incorporated a broad trapeziform slot at

TABLE 1.2

Techniques Used for Obtaining Dual
Band-notched Characteristics

| Sl. No. | Geometry | Reference |
|---------|----------|-----------|
| 1 | C-shaped | [38, 39, 49, 50] |
| 2 | U-shaped | [41,43, 47, 49] |
| 3 | E-shaped | [42, 54] |
| 4 | I/J-shaped | [47, 54] |
| 5 | H-shaped | [44] |
| 6 | L-shaped | [40, 42, 43] |

the ground plane with embedded stubs and inverse trapeziform-shaped fed metal stub. A compact antenna proposed by Azim et al. [55] is investigated with two partial and semi-circular annular slots at the upper and lower side of the ring radiator for achieving dual notched frequency band. Nguyen and Maeda [56] have used via hole structures to connect additional patches placed on the back of the substrate with the radiating patch to realize dual band-notched characteristic. Fractal binary tree-based compact UWB antenna is presented by Jahromi et al. [57] for achieving a dual band-notch characteristic.

Thus, on analyzing the above literatures [38–57], it can be noted that in most of the cases, the alphabetical slots take over the notch characteristics. Table 1.2 represents a brief description of different alphabetical geometries used for notch functioning. The literature is further categorized based on different methods used for obtaining notch characteristics for both microstrip and CPW feeding in Figure 1.4.

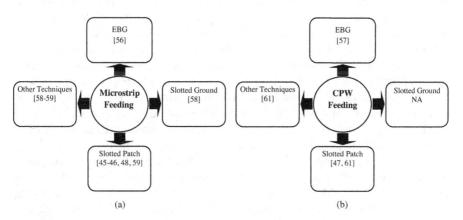

(a)　　　　　　　　　　　　　(b)

FIGURE 1.4
Feeding structures for dual notch characteristics based on (a) microstrip feed and (b) CPW feed.

## 1.5 MultiBand-Notched Characteristics

In order to increase the applicability of UWB antennas and reduce the probability of potential interference from nearby narrow-band communication systems, the UWB antennas with multi-notch characteristics are extensively developed nowadays. By etching multiple slots or installing multiple stubs parasitic elements and EBG structures triple [58–68], quadruple [69–74] notch characteristics are obtained.

As in case of single and dual notch antenna configurations, slotted geometries are also used in order to produce triple band-notch characteristics. U- and L-shaped slots are exerted in [58], whereas U- and C-shaped slots are utilized in [59] for rejecting the undesired frequency bands. The triple band-rejection function is obtained after utilizing concentric G-shaped slots and U-shaped feed line slot in [60]. Nguyen et al. [61] have designed three open-ended quarter wavelength slots to create as many numbers of notches. Liu et al. [62] have obtained band-notch functioning monopole antenna by employing a complimentary split ring resonator (CSRR) slotted radiator along with a slotted feed line. A pair of CLL resonator is used to reject WLAN frequency band in [63] as well as in [64]. For creating notch-band characteristics, EBG structures are also being used in UWB antennas. In [65], triple notched-band property has been obtained by using complementary EBG (CEBG) on both sides of the feed line and ring on the ground. Four different configurations of beveled rectangular patch UWB antenna are designed [66]. Therefore, by utilizing one, two, and three EBG structures, single, dual, and triple band-rejection properties have been obtained at 3.4, 5.2, and 5.8 GHz, respectively. However, to verify the effect of notches, all the structures are simulated individually.

Different other techniques are also employed for acquiring triple notch functioning for designing UWB antennas. T-shaped slots along with loaded T-shaped stub are implemented in [67]. Again, T-shaped stubs are utilized together with an arc-shaped stub in [68] for procuring desired triple notch quality.

Likewise, for attaining quad notched-band characteristics, a compact microstrip-fed UWB elliptical monopole antenna based on four U-shaped slots is proposed for four band-notched characteristics [69]. A simple design of multilayered quad band-notched UWB antenna is presented by adding close loop ring resonators [70]. Li et al. [71] have designed a novel compact UWB monopole antenna with four band-notched characteristics depending on the crescent-shaped resonators and U-shaped slot in the ground. Yuchang et al. [72] have proposed CPW-fed monopole antenna using split ring slots to realize quadruple band-notched characteristics suitable for UWB applications, and then a CPW-fed antenna has been developed based on split ring resonator (SRR) and CSRR structures by Cao et al. [73]. Darvish et al. [74] have presented a compact CPW-fed printed monopole antenna with various configurations of modified-U (MU)-slots for multi-band notches.

**TABLE 1.3**

Techniques Used for Obtaining Multi Band-notched Characteristics at Different Frequencies

| Sl. No. | Technique | Notched Frequency Band | | | |
|---|---|---|---|---|---|
| | | WiMAX | WLAN | X-/ITU Band | Other |
| 1 | SRR/CSRR | [64, 72, 73] | [72, 73] | [72, 73] | NA |
| 2 | U-shaped slot | [58–59, 69, 74] | [60, 69, 71, 74] | [59, 71] | [58, 69, 71, 74] |
| 3 | C-/G-/L-shaped slot | [60–61] | [58–59] | NA | [60] |
| 4 | Capacitively loaded loop resonators | [70] | [63, 70] | [70] | NA |
| 5 | Parasitic segments | NA | NA | NA | [63] |
| 6 | T- and arc-shaped stub/slot | [68] | [68] | [68] | NA |
| 7 | Open-ended quarter wavelength slots/stubs | [63] | [61] | [61] | NA |
| 8 | Other techniques | [64] | NA | NA | NA |

Based on the literature available here [58–68], triple and [69–74] quad band-notch characteristics are realized on different antenna configurations with various techniques proposed at different frequencies. These band-notched techniques are summarized in Table 1.3 with their cited references.

## 1.6 Reconfigurable and Tunable Band-Notched Characteristics

Reconfigurable antennas are highly desirable in recent times for enhancing the performance of UWB systems. These antennas are capable of providing switchable band-notch performance. In their usual state of operation, reconfigurable antennas can utilize the whole UWB spectrum. However, when such antennas encounter an interfering signal, they generate band-notch function by changing their configuration which then eliminates interference with the coexistent system. The reconfigurable [75–84] and tunable [85–90] antenna structures can be achieved by employing different kind of RF switches such as PIN diodes, RF microelectromechanical system (RF MEMS), and varactor diodes.

Valizade et al. [75] have utilized pi-shaped slot integrated with PIN diode for reconfigurable and multi resonance function. Parasitic strip and a T-shaped stub, connected to the radiating stub, have been configured by a PIN diode for achieving antenna reconfigurable capability [76]. With just two PIN diodes, three switchable band-notch characteristics are investigated by Tasouji et al. [77]. Badamchi et al. [78] have embedded slotted geometries into the radiating patch, and then integrated PIN diodes for switching between single and dual notch performances. PIN diode switches are again embedded between U-shaped slots and parasitic patch in [79], and similarly, they are added between arc-shaped slots in [80] for electronically switching the notch frequencies. Majid et al. [81] have used EBG structures connected with PIN diode switches to enable the reconfigurable operation. A reconfigurable band-notched CPW-fed UWB antenna using EBG structure is introduced in [82]. The notched band at 4.0 GHz has been obtained by placing EBG structure adjacent to the transmission line. Furthermore, the band-notched characteristic can be disabled by switching the state of switch place at the strip-line. Nikolau et al. [83] have introduced a CPW-fed compact elliptical monopole UWB antenna, where the reconfigurable frequency characteristics have been realized by switching the inverted L-shaped open stubs. Then, they have proposed two antenna configurations with U-shaped slot and L-shaped stubs, respectively, and using MEMS switches for reconfigurable operations [84].

In order to reduce the interference completely, UWB band-notched antennas with tunability feature are highly desirable. A tunable and reconfigurable circular slot antenna with frequency-rejection characteristics is proposed for UWB communication applications, which can provide triple band-notch characteristics [85]. Wu et al. [86] have proposed band-notched UWB antenna with a novel approach of integrating the switchable and tunable properties. Again, S-shaped split ring resonators loaded with a pair of varactor diodes are allowed to couple to a CPW line for generating the desired notched behavior [87]. Xia et al. [88] have designed dual tunable band-notched printed monopole UWB antenna. The notched-band functions of 3.3–3.7 GHz for WiMAX and 5.15–5.825 GHz for WLAN are obtained by introducing C-shaped slots above and below the substrate layer, and their central notched frequencies are tuned individually by placing a capacitor and an inductor on each of the slots. A UWB antenna with single band-notched function at 5.8 GHz is presented in [89]. The antenna consists of pyramidal radiating patch with slots in each of the faces of pyramid. Therefore, by introducing lumped capacitor or varactor diode over the slots, the notch-band center frequency has been tuned from 4.8 GHz to 7.472 GHz by varying the capacitor value from 0.1 pF to 10 pF. A miniaturized UWB antenna with single tunable band-notched characteristics at 5.5 GHz is presented in [90]. The notched-band function is accomplished by realizing C-shaped slot above the radiating patch.

## 1.7 Band-Notched UWB MIMO Antennas

High-speed and reliable data transmission without an increment in the bandwidth or the transmitted power is very obligatory. To achieve this goal, multiple transmitting and receiving antennas are used to achieve spatial diversity or spatial multiplexing. However, placing multiple elements causes strong coupling between them. In this section, analysis of different MIMO configurations UWB antenna with single, dual, and multiple notch characteristics has been carried out [91–100]. In [91–95], several slotted geometry have been used, whereas parasitic stubs are used in [96–98]. Finally, EBG structures are being used [99, 100] for creating notched-band behavior, but the literature is very limited.

To realize band-notched characteristics at WLAN band, the suggested antenna in [91] has been incorporated with C-shaped slot, and L-shaped ground stub has been used for obtaining better isolations. Moreover, the proposed structure provides miniaturization using Minkowski-shaped fractal geometry. Similarly, by using C-shaped slot and meander line structure, the suggested design produces notched-band function for WLAN applications with –17.5 dB isolation [91]. Lin et al. [92] have designed 4 × 4 UWB MIMO antenna. The antenna comprises rectangular patch with I- and C-shaped slots for realizing dual notched behaviors in the frequency ranges of WiMAX (3.3–3.8 GHz) and WLAN (5.15–5.35 GHz). Therefore, to meet high coupling reduction better than 20 dB, four rectangular and four staircase-shaped stubs are added to the ground plane. Dual band-notched UWB MIMO antenna is designed in [93]. The antenna consists of one-third $\lambda$-open-ended slot, half $\lambda$-parasitic strips, and one-fourth $\lambda$-open-ended slot for realizing notched characteristics in the frequency ranges of WLAN (5.15–5.825 GHz) and WiMAX (3.3–3.7 GHz). Recently, rectangular SRR slots have been used for realizing notched-band function at WiMAX and WLAN bands. The suggested antenna provides an isolation of –15 dB by proposing meander-shaped ground stubs [94]. The analysis of compact UWB monopole antenna with dual diversity MIMO configuration and dual band-notched characteristics has been carried out in [95]. The antenna consists of U-shaped slot and horizontal line slot over the pacman-shaped radiator to reject interference within WiMAX (3.5 GHz) and WLAN (5.5 GHz).

For creating band-notched characteristics in UWB antennas, loading of parasitic elements or stubs is also a very common technique. However, this technique makes design complex. Dual band-notch MIMO antenna with two U-shape patches has been evolved for UWB applications [96]. Thus, by introducing trekking slots in the patch and loading C-shaped strips around the feed line, notched-band characteristics are generated within WiMAX (3.3–3.7 GHz) and WLAN (5.2–5.8 GHz) bands. Finally, to reduce the mutual coupling between elements, a fork-shaped stub is loaded in the ground plane. A single band-notched UWB MIMO antenna is presented in [97]. The antenna comprises two radiators separated by an isolated element to reduce the mutual coupling between them. For creating notched function, parasitic slits are placed on both the elements. Moreover, the design has good performance characteristics in terms of improving gain and radiation patterns, with envelope correlation coefficient (ECC) less than about 0.07. Dual polarized band rejection UWB MIMO antenna is designed by Zhu et al. [98].

Two quasi-self-complementary (QSC) antenna patches are being used for obtaining polarization diversity and better isolation. Notched band at WLAN system has been realized by etching a bent slit in each of the radiating elements. Further, the authors have modified the design structure with four-element MIMO system to observe the performance behavior.

Recently, EBG structures are being used for generating notch characteristics in UWB antennas. Besides this, they are also being used for reduction in mutual coupling in antenna arrays and improvement in gain. Literatures [99, 100] have used various EBG structures for generating band-notched functions in UWB antennas. A single band-notched UWB MIMO antenna is designed using modified defected ground plane and a periodic EBG structure [99]. The proposed antenna exhibits impedance bandwidth of 3–16.2 GHz and is having a sharp band-notched at 4.6 GHz. Besides this, an isolation of 17.5 dB has been achieved with a peak gain of 8.4 dB. A pair of mushroom-type EBG structure has been used over circular monopole UWB MIMO antenna for realizing band-notched function at 5.8 GHz. Moreover, the suggested structure provides an isolation of 20 dB [100].

After analyzing the above literatures [91–100], a categorization of the different methods applied for obtaining notch characteristics in MIMO antenna configurations is presented in this section.

## 1.8 Summary

In this chapter, the literature on UWB antennas with notch characteristics has been characterized into different sections covering around 100 research articles. Various types of UWB antennas, such as monopole antennas, tapered slot antennas, and Vivaldi antennas, are considered for analysis in this chapter. For ease of smooth understanding, these are classified based on their notch properties. It has been observed that almost all antennas cover the frequency band between 3.1 GHz and 10.6 GHz along with single or multiple notch characteristics. The operation over the UWB range is mainly due to the radiating patches of different geometries, primarily rectangular, circular, elliptical, or other modified shapes. The notches can be attributed to incorporation of different types of slots either on the radiating patch or on the ground plane of the antennas. Addition of parasitic elements or open-ended stubs has also turned out to be an effective way for procuring the desired notch behavior. In addition to that, EBG structures are also being used for generating notched-band characteristics.

From the above sections, we can observe that for single notch case, the band rejection is targeted for WLAN applications in almost all the antenna configurations. For dual notch configurations, almost all band rejections are targeted for centering at ~3.5 GHz and ~5.5 GHz, with a few bands at other frequencies. Similarly, for multi-notch antennas, focus has been on rejecting WiMAX, WLAN, and X-band frequencies in most of the configurations. However, some of the proposed configurations have also been included with notch reconfigurable and tunable properties using PIN diodes, MEMS switch, and varactor diode. Moreover, for better understanding, each section is described in detail in Chapters 2–7.

## 1.9 Conclusion

This chapter has addressed performance of different notch antennas designed for application in UWB systems, and for their ability to evade interferences from the existing narrow-band communication systems. Almost all the major techniques that are employed for acquiring notch behavior are highlighted and comprehensively addressed in this chapter. It has been observed that the notch antennas designed for rejecting frequency bands like X-band, C-band, ISM band, etc. are very less in the literature. Also, notch antennas with quadruple characteristics, reconfigurable, and tunable ability are less in literature as compared to single/dual/and triple notch antennas. Further, it has also been observed that literature on UWB MIMO antennas using EBG structure is very limited which is beneficial for the researcher to work in that domain.

## References

[1] J. Liang, Antenna Study and Design for Ultra Wideband Communication Applications, United Kingdom: University of London, 2006.
[2] Federal Communications Commission, "Revision of part 15 of the commission's rules regarding ultra-wideband transmission systems," Tech. Rep. ET-Docket 98-153, FCC02-48, Federal Communications Commission (FCC), Washington, DC, USA, 2002.
[3] H. G. Schantz, "Introduction to Ultra-Wide Band Antennas," In: Proceedings of 2003 IEEE Conference on Ultra Wideband System Technologies, Reston, USA, pp. 1–9, November, 2003.
[4] A. Stark, C. Friesicke, J. Muller, A. F. Jacob, "A packaged ultrawide band filter with high stopband rejection," IEEE Microwave Magazine, vol. 11, no. 5, pp. 110–117, 2010.
[5] S. T. Fan, Y. Z. Yin, H. Li, L. Kang, "A novel self-similar antenna for UWB applications with band-notched characteristics," Progress in Electromagnetics Research Letters, vol. 22, pp. 1–8, 2011.
[6] A. Dastranj, "Optimization of a printed UWB antenna: Application of the invasive weed optimization algorithm in antenna design" IEEE Antennas and Propagation Magazine, vol. 59, no. 1, pp. 48–57, 2017.

[7] Y. Kim, D. H. Kwon, "CPW-fed planar ultra wideband antenna having a frequency band notch function," Electronics Letters, vol. 40, no. 7, pp. 403–405, 2004.

[8] Y. Kim, D. H. Kwon, "Planar Ultra Wide Band Slot Antenna with Frequency Band Notch Function," In: Proceedings of IEEE Antennas and Propagation Society Symposium, USA, vol. 2, pp. 1788–1791, June, 2004.

[9] Z. Li, Y. Wang, J. Wang, K. Jing, "An Ultra-Wideband Planar Monopole Antenna with Band-Notched Characteristics," In: Proceedings of 2009 International Conference on Microwave Technology and Computational Electromagnetics (ICMTCE 2009), China, pp. 62–65, November, 2009.

[10] S. Nikolaou, B. Kim, Y. S. Kim, J. Papapolymerou, M. M. Tentzeris, "CPW-Fed Ultra Wideband (UWB) Monopoles with Band Rejection Characteristic on Ultra Thin Organic Substrate," In: Proceedings of Asia-Pacific Microwave Conference 2006 (APMC), Japan, December, 2006.

[11] M. Ojaroudi, G. Ghanbari, N. Ojaroudi, C. Ghobadi, "Small square monopole antenna for UWB applications with variable frequency band-notch function," IEEE Antennas and Wireless Propagation Letters, vol. 8, pp. 1061–1064, 2009.

[12] Y. Hacene, X. Shuguo, T. Rahman, "Design of A Novel Monopole Antenna with 5.5 GHz Band-Notch Characterization for UWB Applications," In: Proceedings of 10th International Symposium on Antennas, Propagation & EM Theory (ISAPE), China, pp. 273–276, October, 2012.

[13] A. T. Norzaniza, M. A. Matin, "Design of Microstrip UWB Antenna with Band Notch Characteristics," In: Proceedings of 2013 IEEE TENCON Spring Conference, Sydney, Australia, pp. 51–52, April, 2013.

[14] M. A. Matin, M. M. Hossain, "A New Planar Printed Antenna with Band-Notch Characteristics for UWB Applications," In: Proceedings of 2015 IEEE Region 10 Conference (TENCON), China, pp. 1–3, November, 2015.

[15] Q. X. Chu, Y. Y. Yang, "A Compact CPW-Fed Planar Ultra-Wideband Antenna with A Frequency Notch Characteristic," In: Proceedings of Asia-Pacific Microwave Conference (APMC 2007), Thailand, pp. 1–4, December 2007.

[16] Y. Y. Yang, Q. X. Chu, Z. A. Zheng, "Time domain characteristics of band-notched ultra-wideband antenna," IEEE Trans. on Antennas and Propagation, vol. 57, no. 10, pp. 3426–3430, 2009.

[17] Y. D. Dong, W. Hong, Z. Q. Kuai, J. X. Chen, "Analysis of planar ultra-wideband antennas with on-ground slot band-notched structures," IEEE Transactions on Antennas and Propagation, vol. 57, no. 7, 2009.

[18] P. Rakluea, J. Nakasuwan, "Planar UWB Antenna with Single Band-Notched Characteristic," In: Proceedings of 2010 International Conference on Control, Automation and Systems (ICCAS), South Korea, pp. 1978–1981, October, 2010.

[19] Y. F. Weng, W. J. Lu, S. W. Cheung, T. I. Yuk, "UWB Antenna with Single or Dual Band-Notched Characteristic for WLAN Band Using Meandered Ground Stubs," In: Proceedings of Antennas & Propagation Conference, United Kingdom, pp. 757–760, November, 2009.

[20] A. Nouri, G. R. Dadashzadeh, "A compact UWB band-notched printed monopole antenna with defected ground structure," IEEE Antennas and Wireless Propagation Letters, vol. 10, pp. 1178–1181, 2011.

[21] B. Biswas, D. R. Poddar, R. Ghatak, A. Karmakar, "Modified Sierpinski Carpet Fractal Shaped Slotted UWB Monopole Antenna with Band Notch Characteristic," In: Proceedings of National Conference on Communications (NCC), India, pp. 1–5, February, 2013.

[22] L. Yao, J. Xia, H. Zhu, N. Li, X. Li, "A High Gain UWB Vivaldi Antenna with Band Notched Using Capacitively Loaded Loop (CLL) Resonators," In: Proceedings of International Conference on Microwave and Millimeter Wave Technology (ICMMT), China, vol. 2, pp. 820–822, June, 2016.

[23] M. A. S. Al-zahrani et al., "Design and Performance Analysis of An Ultra-Wideband Monopole Microstrip Patch Antenna with Enhanced Bandwidth and Single Band-Notched Characteristics," In: Proceedings of 2017 Progress in Electromagnetics Research Symposium-Fall (PIERS-FALL), Singapore, pp. 1488–1493, November, 2017.

[24] N. H. M. Sobli, H. E. Abd-El-Raouf, "Design of A Compact Band-Notched Antenna for Ultrawideband Communication," In: Proceedings of IEEE Antennas and Propagation Society International Symposium, USA, pp. 1–4, July, 2008.

[25] K. Ruchandani, M. Kumar, "A Novel CPW Fed Octagonal Aperture UWB Antenna with 5GHz/6GHz Band Rejection by H Shaped Slot," In: Proceedings of International Conference on Computer, Communication and Control (IC4), India, pp. 1–4, September, 2015.

[26] A. A. Kalteh, R. Fallahi, M. G. Roozbahani, "A Novel Microstrip-Fed UWB Circular Slot Antenna with 5-GHz Band-Notch Characteristics," In: Proceedings of the IEEE International Conference on Ultra-wideband, Germany, vol. 1, pp. 117–120, September, 2008.

[27] R. Movahedinia, M. N. Azarmanesh,"Ultra-wideband band-notched printed monopole antenna," IET Microwave Antennas Propagation, vol. 4, no. 12, pp. 2179–2186, 2010.

[28] S. Hong, K. Chung, J. Choi, "Design of A Band-Notched Wideband Antenna for UWB Applications," In: Proceedings of 2006 IEEE Antennas and Propagation Society International Symposium, USA, pp. 1709–1712, July, 2006.

[29] H. K. Yoon et al., "UWB Wide Slot Antenna with Band-Notch Function," In: Proceedings of 2006 IEEE Antennas and Propagation Society International Symposium, USA, pp. 3059–3062, July, 2006.

[30] S. Ghosh, "Band-notched modified circular ring monopole antenna for ultra wideband applications," IEEE Antennas and Wireless Propagation Letters, vol. 9, pp. 276–279, 2010.

[31] Y. S. Li, X. D. Yang, C. Y. Liu, T. Jiang, "Compact CPW-fed ultra-wideband antenna with band-notched characteristic," Electronics Letters, vol. 46, no. 23, pp. 1533–1534, 2010.

[32] M. Yazdi, N. Komjani, "Design of a band-notched UWB monopole antenna by means of an EBG structure," IEEE Antennas and Wireless Propagation Letters, vol. 10, pp. 170–173, 2011.

[33] N. Jaglan, S. Dev Gupta, "Reflection phase characteristics of EBG structures and WLAN band notched circular monopole antenna design," International Journal on Communications Antenna and Propagation (IRECAP), pp. 233–240, August, 2015.

[34] K. Chang, H. Kim, Y. J. Yoon, "Multi-Resonance UWB Antenna with Improved Band Notch Characteristics," In: Proceedings of 2005 IEEE Antennas and Propagation Society International Symposium, USA, vol. 3A, July, 2005.

[35] T. Nakamura, H. Iwasaki, "Planar Monopole Antenna Having A Band-Notched Characteristic for UWB," In: Proceedings of IEEE Antennas and Propagation Society International Symposium, USA, pp. 4641–4644, July, 2006.

[36] Q. X. Chu, M. H. Qing, H. Zheng, "Design of shared aperture wideband antennas considering band-notch and radiation pattern control," IEEE Transactions on Antennas Propagation, vol. 56, no. 11, 2008.

[37] A. Ghobadi, C. Ghobadi, J. Nourinia, "A novel band-notched planar monopole antenna for ultra-wideband applications," IEEE Antennas and Wireless Propagation Letters, vol. 9, pp. 608–611, 2010.

[38] Q. X. Chu, Y. Y. Yang, "3.5/5.5 GHz dual band-notch ultra-wideband antenna," Electronics Letters, vol. 44, no. 3, pp. 172–174, 2008.

[39] X. Gao et al., "CPW-Fed Slot Antenna with Dual Band-Notched Characteristic for UWB Application," In: Proceedings of 2012 Fourth International High Speed Intelligent Communication Forum, China, pp. 1–3, May 2012.

[40] D. Yu, G. Xie, Z. Liao, W. Zhai, "Novel Dual Band-Notched Omni-Directional UWB Antenna on Double Substrates Crossing," In: Proceedings of 2011 Fourth IEEE International Symposium on Microwave, Antenna, Propagation and EMC Technologies for Wireless Communications, China, pp. 95–97, November, 2011.

[41] J. Wu, J. Li, "Compact Ultra-Wideband Antenna with 3.5/5.5GHz Dual Band-Notched Characteristic," In: Proceedings of 2013 Fifth IEEE International Symposium on Microwave, Antenna, Propagation and EMC Technologies for Wireless Communications, China, pp. 446–450, October, 2013.

[42] M. Mehranpour, J. Nourinia, C. Ghobadi, M. Ojaroudi, "Dual band-notched square monopole antenna for ultrawideband applications," IEEE Antennas and Wireless Propagation Letters, vol. 11, pp. 172–175, 2012.

[43] A. Yadav, D. Sethi, S. Kumar, S. L. Gurjar, "L and U Slot Loaded UWB Microstrip Antenna: C-band/WLAN Notched," In: Proceedings of 2015 IEEE International Conference on Computational Intelligence & Communication Technology (CICT), India, pp. 380–384, February 2015.

[44] Y. E. Jalil, C. K. Chakrabarty, B. Kasi, "A Compact Ultra Wideband Antenna with Dual Band-Notched Design," In: Proceedings of 2013 Seventh International Conference on Signal Processing and Communication Systems, Australia, pp. 1–5, December 2013.

[45] F. Zhao et al., "Design of Novel Dual Band-Notched Disk Monopole Antennas," In: Proceedings of The 2012 International Workshop on Microwave and Millimeter Wave Circuits and System Technology (MMWCST), China, pp. 1–4, April, 2012.

[46] A. Das, S. P. Singh, S. Sahu, "Compact Microstrip Fed UWB Antenna with Dual Band Notch Characteristics," In: Proceedings of 2016 International Conference on Communication and Signal Processing (ICCSP), India, pp. 751–754, April, 2016.

[47] J. Y. Zhao, G. Fu, L. Y. Ji, Q. Y. Lu, Z. Y. Zhang, "Compact printed Ultra-Wideband Antenna with Dual Band-Notched Characteristics," In: Proceedings of 2011 International Conference on Electronics, Communications and Control (ICECC), China, pp. 780–782, September 2011.

[48] W. Jiang, W. Che, "A novel UWB antenna with dual notched bands for WiMAX and WLAN applications," IEEE Antennas and Wireless Propagation Letters, vol. 11, pp. 293–296, 2012.

[49] J. Shen, C. Lu, J. Zhang, "Heart-Shaped Dual Band-Notched UWB Antenna," In: Proceedings of 2014 Third Asia-Pacific Conference on Antennas and Propagation, China, pp. 487–490, July, 2014.

[50] N. Ojaroudi, M. Ojaroudi, "Novel design of dual band-notched monopole antenna with bandwidth enhancement for UWB applications," IEEE Antennas and Wireless Propagation Letters, vol. 12, pp. 698–701, 2013.

[51] J. D. Guang, F. Jiayue, Z. X. Wang, "Dual Band-Notched UWB Antenna with Folded SIRs," In: Proceedings of 2012 International Conference on Microwave and Millimeter Wave Technology (ICMMT), China, vol. 3, pp. 1–3, May, 2012.

[52] Y. Wang, T. Huang, D. Ma, P. Shen, J. Hu, W. Wu, "Ultra-Wideband (UWB) Monopole Antenna with Dual Notched Bands by Combining Electromagnetic-Bandgap (EBG) and Slot Structures," IEEE MTT-S International Microwave Biomedical Conference (IMBioC), Nanjing, China, pp. 1–3, 2019.

[53] M. Ghahremani, C. Ghobadi, J. Nourinia, M. S. Ellis, F. Alizadeh, B. Mohammadi, "Miniaturised UWB antenna with dual-band rejection of WLAN/WiMAX using slitted EBG structure," IET Microwaves, Antennas & Propagation, vol. 13, no. 3, pp. 360–366, 27 2 2019.

[54] L. Q. Kun, X. Feng, Z. G. Qiu, T. Zhen, "Design of A Planar Ultra-Wideband Antenna with Dual Band-Notched Characteristics," In: Proceedings of 2010 IEEE 12th International Conference on Communication Technology (ICCT), China, pp. 179–182, November 2010.

[55] R. Azim, M. S. Alam, N. Misran, A. T. Mobashsher, M. T. Islam, "Compact Planar Antenna with Dual Band-Notched Characteristics for UWB Applications," In: Proceedings of the 2011 IEEE International Conference on Space Science and Communication (IconSpace), Malaysia, pp. 269–272, July, 2011.

[56] H. Nguyen, T. Maeda, "A Compact UWB Antenna with Dual Band-Notched Characteristics Using Viahole Structure," In: Proceedings of The 2011 International Conference on Advanced Technologies for Communications (ATC 2011), Vietnam, pp. 279–282, August, 2011.

[57] M. N. Jahromi, A. Falahati, R. M. Edwards, "Application of fractal binary tree slot to design and construct a dual band-notch CPW-ground fed ultra-wide band antenna," IET Microwave Antennas Propagation, vol. 5, no. 12, pp. 1424–1430, 2011.

[58] S. Maiti, N. Pani, A. Mukherjee, "Modal Analysis and Design A Planar Elliptical Shaped UWB Antenna with Triple Band Notch Characteristics," In: Proceedings of 2014 International conference on Signal Propagation and Computer (ICSPCT), India, pp. 13–15, July, 2014.

[59] S. Tomar, A. Kumar, "Design of A Novel Compact Planar Monopole UWB Antenna with Triple Band-Notched Characteristics," In: Proceedings of 2015 Second International Conference on Signal Processing and Integrated Networks (SPIN), India, pp. 56–59, February, 2015.

[60] B. C. Reddy, E. S. Shajahan, M. S. Bhat, "Design of A Triple Band-Notched Circular Monopole Antenna for UWB Applications," In: Proceedings of 2014 International Conference on Wireless and Optical communications Networks (WOCN), India, pp. 1–5, September, 2014.

[61] D. T. Nguyen, D. H. Lee, H. C. Park, "Very compact printed triple band-notched UWB antenna with quarter-wavelength slots," IEEE Antennas and Wireless Propagation Letters, vol. 11, pp. 411–414, 2012.

[62] W. Liu, T. Jiang, "Design and analysis of a tri-band notch UWB monopole antenna," 2016 Progress In Electromagnetic Research Symposium (PIERS), China, pp. 2039–2041, August, 2016.

[63] S. Nikolaou, M. Davidoviu, M. Nikoliu, P. Vryonides, "Triple Notch UWB Antenna Controlled by Three Types of Resonators," In: Proceedings of 2011 IEEE International Symposium on Antennas and Propagation (APSURSI), USA, pp. 1478–1481, July, 2011.

[64] X. Hu, X. Yang, "Tri-Band-Notched Ultrawideband (UWB) Antenna Using Split Semi-Circular Resonator (SSR) and Capacitvely Loaded Loops (CLL)," In: Proceedings of 2015 IEEE International Conference on Communication Problem Solving (ICCP), China, pp. 186–189, October, 2015.

[65] F. Zhou, Z. Qian, J. Han, C. Peng, "Ultra-Wideband Planar Monopole Antenna with Triple Band-Notched Characteristics," In: International Conference on Microwave and Millimeter Wave Technology, Chengdu, pp. 438–440, 2010.

[66] L. Peng, C. L. Ruan, "Design and time-domain analysis of compact multi-band-notched UWB antennas with EBG structures," Progress in Electromagnetics Research B, vol. 47, pp. 339–357, 2013.

[67] Y. Li, S. Chang, M. Li, X. Yang, "A Compact Ring UWB Antenna with Tri-Notch Band Characteristics Using Slots and Tuning Stub," In: Proceedings of 2011 Fourth IEEE International Symposium on Microwave, Antenna, Propagation and EMC Technologies for Wireless Communications, China, pp. 12–15, November, 2011.

[68] W. Zhang, Y. Li, W. Yu, Y. Dai, "Design and Analysis of An Ultra-Wideband Antenna with Triple Frequency Filtering Characteristics," In: Proceedings of 2015 31st International Review of Progress in Applied Computational Electromagnetics (ACES), Williamsburg, USA, pp. 1–2, May, 2015.

[69] S. H. Z. Deen, R. A. Essa, S. M. M. Ibrahem, "Ultrawideband Printed Elliptical Monopole Antenna with Four Band-Notch Characteristics," In: Proc. of 2010 IEEE Antennas and Propagation Society International Symposium, Toronto, Canada, pp. 1–4, July, 2010.

[70] M. J. Almalkawi, V. K. Devabhaktuni, "Quad band-notched UWB antenna compatible with WiMAX/INSAT/lower-upper WLAN applications," Electronics Letters, vol. 47, no. 19, pp. 1062–1063, 2011.

[71] L. Li, Z. L. Zhou, J. S. Hong, "Compact UWB antenna with four band-notches for UWB applications," Electronics Letters, vol. 47, no. 22, pp. 1211–1212, 2011.

[72] Y. Yang, F. Chun, Y. Sun, H. Yang, "CPW-Fed Monopole Antenna with Quadruple Band-Notched for UWB Application," In: Proceedings of 2011 Cross Strait Quad-Regional Radio Science and Wireless Technology Conference (CSQRWC), Harbin, China, pp. 404–406, July, 2011.

[73] Y. Cao, J. Wu, H. Yang, "Design of CPW-Fed Monopole Antenna with Quadruple Band-Notched Function for UWB Application," In: Proceedings of 2011 International Conference on computational Problem Solving (ICCP), Chengdu, China, pp. 353–356, October, 2011.

[74] M. Darvish, H. R. Hassani, B. Rahmati, "Compact CPW-Fed Ultra Wideband Printed Monopole Antenna with Multi Notch Bands," In: Proceedings of 20th Iranian Conference on Electrical Engineering (ICEE2012), Tehran, Iran, May 2012, pp. 1114–1119.

[75] A. Valizade, C. Ghobadi, J. Nourinia, M. Ojaroudi, "A novel design of reconfigurable slot antenna with switchable band notch and multi resonance functions for UWB applications," IEEE Antennas Wireless Propagation Letters, vol. 11, pp. 1166–1169, 2012.

[76] A. A. Kalteh, G. R. D. Zadeh, M. N. Moghadasi, B. S. Virdee, "Ultra-wideband circular slot antenna with reconfigurable notch band function," IET Microwave Antennas Propagation, vol. 6, pp. 108–112, 2012.
[77] N. Tasouji, J. Nourinia, C. Ghobadi, F. Tofigh, "A novel printed UWB slot antenna with reconfigurable band-notch characteristics," IEEE Antennas Wireless Propagation Letters, vol. 12, pp. 922–925, 2013.
[78] B. Badamchi, J. Nourinia, C. Ghobadi, A. V. Shahmirzadi, "Design of compact reconfigurable ultra-wideband slot antenna with switchable single/dual band notch functions," IET Microwave Antennas Propagation, vol. 8, no. 8, pp. 541–548, 2014.
[79] H. Boudaghi, J. Pourahmadazar, S. A. Aghdam, "Compact UWB Monopole Antenna with Reconfigurable Band Notches Using PIN Diode Switches," In: Proc. of 2013 IEEE Antennas Propag. Society Int. Symposium (APSURSI), Orlando, USA, pp. 1758–1759, July, 2013.
[80] Y. Song et al., "A Compact UWB Antenna with Reconfigurable Band-Notched Characteristics," In: Proceedings of 2014 Third Asia-Pacific Conf. on Antennas Propagation (APCAP), China, pp. 123–126, July, 2014.
[81] H. A. Majid et al., "Wideband antenna with reconfigurable band notched using EBG structure," Progress In Electromagnetics Research Letters, vol. 54, pp. 7–13, June, 2015.
[82] H. A. Majid, M. K. A. Rahim, M. R. Hamid, "Band-notched reconfigurable CPW-fed UWB antenna," Applied Physics, vol. 122, p. 347, 2016.
[83] S. Nikolaou, A. Amadjikpe, J. Papapolymerou, M. M. Tentzeris, "Compact Ultra Wideband (UWB) Elliptical Monopole with Potentially Reconfigurable Band Rejection Characteristic," In: Proceedings of 2007 Asia-Pacific Microwave Conference, Bangkok, Thailand, pp. 1–4, December, 2007.
[84] S. Nikolaou, N. D. Kingsley, G. E. Ponchak, J. Papapolymerou, M. M. Tentzeris, "UWB elliptical monopoles with a reconfigurable band notch using MEMS switches actuated without bias lines," IEEE Transactions on Antennas Propagation, vol. 57, no. 8, pp. 2242–2251, 2009.
[85] Y. Li, R. Mittra, "Tunable and Reconfigurable Frequency Rejection Circular Slot Antenna for UWB Communication Applications," In: PIERS Proceedings, Czech Republic, July 6–9, 2015.
[86] W. Wu, Y. B. Li, R. Y. Wu, C. B. Shi, T. J. Cui, "Band-notched UWB antenna with switchable and tunable performance," International Journal of Antennas and Propagation, vol. 2016, Article ID 9612987.
[87] A. K. Horestani et al., "Reconfigurable and tunable S-shaped split ring resonators and application in band-notched UWB antennas", IEEE Transactions on Antennas Propagation, vol. 64, no. 9, pp. 3766–3776, 2016.
[88] Y. Q. Xia, J. Luo, D. J. Edwards, "Novel miniature printed monopole antenna with dual tunable band-notched characteristics for UWB applications, Journal of Electromagnetic Waves and Applications, pp. 1783–1793, 2010.
[89] Z. H. Hu, P. S. Hall, J. R. Kelly, P. Gardner, "UWB pyramidal monopole antenna with wide tunable band-notched behavior," Electronics Letters, vol. 46, no. 24, pp. 1588–1590, 2010.
[90] A. M. A. Salem, S. I. Shams, A. M. M. A. Allam, "A Miniaturized Ultra Wideband Antenna with Single Tunable Band-Notched Characteristics," In: Asia-Pacific Microwave Conference, Yokohama, pp. 746–749, 2010.

[91] Y. Liu, C. Sun, "A Compact Printed MIMO Antenna for UWB Application with WLAN Band-Rejected," In: International Symposium on Antennas, Propagation and EM Theory (ISAPE), Guilin, pp. 95–97, 2016.

[92] M. Lin, Z. Li, "A Compact $4 \times 4$ Dual Band-Notched UWB MIMO Antenna with High Isolation," In: IEEE International Symposium on Microwave, Antenna, Propagation, and EMC Technologies (MAPE), Shanghai, pp. 126–128, 2015.

[93] J. Li, D. Wu, Y. Wu, G. Zhang, "Dual Band-Notched UWB MIMO Antenna," In: IEEE Asia-Pacific Conference on Antennas and Propagation (APCAP), Kuta, pp. 25–26, 2015.

[94] K. Chhabilwad, G. S. Reddy, A. Kamma, B. Majumder, J. Mukherjee, "Compact Dual Band Notched Printed UWB MIMO Antenna with Pattern Diversity," In: IEEE International Symposium on Antennas and Propagation & USNC/URSI National Radio Science Meeting, Vancouver, British Columbia, pp. 2307–2308, 2015.

[95] S. Naser, N. Dib, "A Compact Printed UWB Pacman-Shaped MIMO Antenna with Two Frequency Rejection Bands," In: IEEE Jordan Conference on Applied Electrical Engineering and Computing Technologies (AEECT), Amman, pp. 1–6, 2015.

[96] D. D. Katre, R. P. Labade, "Higher Isolated Dual Band Notched UWB MIMO Antenna with Fork Stub," In: IEEE Bombay Section Symposium (IBSS), Mumbai, pp. 1–5, 2015.

[97] K. Lin, L. Hwang, C. Hsu, S. Wang, F. Chang, "A Compact Printed UWB MIMO Antenna with a 5.8 GHz Band Notch," In: International Symposium on Antennas and Propagation Conference Proceedings, Kaohsiung, pp. 419–420, 2014.

[98] J. Zhu, S. Li, B. Feng, L. Deng, S. Yin, "Compact dual-polarized UWB quasi-self-complementary MIMO/diversity antenna with band-rejection capability," IEEE Antennas and Wireless Propagation Letters, vol. 15, p. 905–908, 2016.

[99] W. Wu, B. Yuan, A. Wu, "A quad-element UWB-MIMO antenna with band-notch and reduced mutual coupling based on EBG structures," International Journal of Antennas and Propagation, vol. 2018, Article ID 8490740, 10, 2018.

[100] D. Z. Nazif, R. S. Rabie, M. A. Abdalla, "Mutual Coupling Reduction in Two Elements UWB Notch Antenna System," In: IEEE International Symposium on Antennas and Propagation & USNC/URSI National Radio Science Meeting, San Diego, CA, pp. 1887–1888, 2017.

[101] H. G. Schantz, G. Wolenec, E. M. Myszka, "Frequency Notched UWB Antennas," In: Proceedings of the 2003 IEEE Conference on Ultra Wideband Systems and Technologies, Reston, USA, pp. 214–218, November, 2003.

[102] J. Xu et al., "A small UWB antenna with dual band-notched characteristics," International Journal of Antennas and Propagation, vol. 2012, Article ID 656858.

[103] W. S. Lee, W. G. Lim, J. W. Yu, "Multiple band-notched planar monopole antenna for multiband wireless systems," IEEE Microwave and Wireless Components Letters, vol. 15, no. 9, pp. 576–578, 2005.

[104] D. Guha, Y. M. M. Antar, Microstrip and Printed Antennas: New Trends, Techniques and Applications, United Kingdom, Wiley, 2010.

# 2

# Single Band-Notched UWB Antennas

## 2.1 Introduction

Over recent years, the demand for ultra-wideband communication systems is rapidly growing after the release of unlicensed frequency bands 3.1–10.6 GHz, by the federal communication commission (FCC) for commercial applications. Since then, it has become a promising technology for short-range communications with wider bandwidth and low power consumption [1, 2]. Moreover, in recent days, UWB technology has gained a lot of attention among the researchers as well as the wireless communication industries. Traditionally, it has been used for radar imaging but recently the technological enhancements have allowed UWB to be implemented for new services related to mobile communication such as voice, audio, video, and data services. Further, this technology has also helped us in achieving faster exchange of data between portable devices [3]. As an essential part of communication systems, antennas have drawn heavy attention from academicians. Currently, there is an increased interest in developing different antenna configurations for use in several applications like satellite communication, space application, and wireless communication [4]. Microstrip antennas are the current day antenna designer's choice because of their numerous advantages such as compact size, low profile, low fabrication cost, light-weight characteristics, and easy integration with microwave circuits [5, 6]. Although microstrip patch antennas are compact in size, they have a narrow impedance bandwidth, usually less than 5% [7]. Therefore, it is necessary to design an antenna with wide impedance bandwidth and compact size. However, designing of antenna for UWB communication imposes several challenge, including impedance matching, radiation stability, compact size, etc. [8]. Besides these features, UWB antenna faces a major challenge on band interfering issues. Various narrow bands like WiMAX band (3.3–3.7 GHz), WLAN band (5.15–5.35 GHz and 5.725–5.825 GHz), C-band satellite bands (3.77–4.2 GHz downlink band, 5.9–6.4 GHz uplink band), and X-band (8.025–8.4 GHz) already exist within the UWB region and cause overlapping of spectrum [9–12]. Therefore, to overcome the interfering issues, various techniques have been adopted till date which are discussed further.

In this chapter, a review has been carried out on characteristics of single band notched UWB antennas and the various techniques for introducing notches in UWB systems. Single band-notched UWB antennas are commonly realized using techniques such as embedding different types of slots on the radiating patch or ground plane or the feeding line [13–38]. Besides, by introducing stub-type elements or parasitic elements [39–51], single notched band behavior has been retained. Moreover, electromagnetic bandgap (EBG) structures [52–56] are also being recently used to reject the potential interference within the specified UWB frequency range.

## 2.2  Slots-Loaded Geometries

A planar UWB antenna with single band-notched characteristic for rejecting interference from the WLAN band is presented in [13]. The notch at 4.8–5.5 GHz is obtained by embedding a rectangular slot on the ground plane. A CPW-fed planar antenna for UWB application is presented in [14]. The proposed antenna consists of a beveled rectangular patch and a modified rectangular slotted ground. By cutting a half-wave resonant structure in the radiating patch, a notch band centered at 5.6 GHz for band rejection at WLAN is achieved. Moreover, the center frequency and bandwidth of the notch band can easily be adjusted by varying the parameters of the resonant slot. In [15], an ultra-wideband truncated rectangular monopole antenna with band-notched characteristics is presented and shown in Figure 2.1(a). The antenna consists of a split ring resonator (SRR) that is etched inside the truncated rectangular patch of the monopole antenna to achieve band-notch for WLAN operation and has an impedance bandwidth of 2.43–13.30 GHz. Kim and Kwon [16] have proposed a single notched band CPW-fed UWB planar antenna using V-shaped slot. Nikolaou et al. [17] have designed an elliptical monopole UWB antenna using liquid crystal polymer substrate and having notched band function within 5.8 GHz using U- and C-shaped slot over antenna radiator. A UWB antenna with high resolution and high data transmission rates is proposed in [18]. Thus, by inserting a thin V-shaped slot, a narrow frequency band-notch function has been obtained. In [19], a circular slot-type SRR is used for creating notched band characteristics at the 5.2 GHz, whose geometrical configuration is depicted in Figure 2.1(b) and its performance parameters are shown in Figure 2.2. A CPW-fed planar antenna for ultra-wideband application with band-notch characteristics is presented in [20]. The proposed antenna consists of a rectangular metal patch embedded with a C-shaped slot for creating notch at 5–6 GHz for the WLAN band. The proposed antenna yields an impedance bandwidth of 3.1–10.6 GHz. Similarly, by using L-shaped slot [21] and two slits [22] on the radiating patch (Figure 2.1(c)), a notched band function within the WLAN band is obtained. A square metal-plate monopole UWB antenna with band-notched property is presented in [23]. The antenna consists of a square metal plate with two bevel-cuts, an inverted V-shaped slot for rejecting EM interference at 5–6 GHz. A CPW-fed UWB-printed circular monopole antenna with SRR for achieving band-notch characteristics in the 5.15–5.825 GHz band

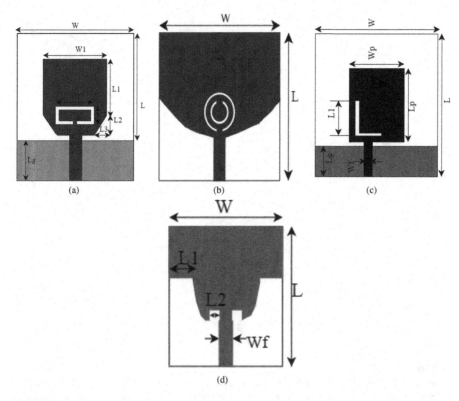

**FIGURE 2.1**
Different types of slot loaded single band notched UWB antennas [re-drawn] (a) SRR loaded antenna [15], (b) Circular SRR loaded antenna [19], (c) L-shaped slot antenna [21], and (d) Shovel-shaped DGS-antenna [25].

is presented in [24]. Nouri and Dadashzadeh [25] have proposed a simple and compact UWB-printed monopole antenna (PMA) with filtering characteristics. The filtering property at WLAN band has been achieved by using a modified shovel-shaped DGS as presented in Figure 2.1(d).

Figure 2.3(a) and (b) reveals that PMAs with band rejection characteristics at WLAN band can be realized by using two modified rectangular slots, as introduced in [26], and also by segmenting a circular monopole patch [27]. Likewise, by inserting arc-shaped slot above the radiating patch sharp band-notch characteristics from 4.9 GHz to 5.92 GHz can be accomplished [28]. Mahmood et al. [29] have proposed a compact band-notch UWB antenna with a CPW-fed system. The notched characteristics at 5–5.7 GHz are designed by etching a hexagonal-shaped slot in the radiation element. A low-cost compact microstrip band-notch UWB antenna is designed and presented in [30]. The notched band characteristics within WLAN band (5.1–5.8 GHz) is retained by using U-shaped slot in the radiating patch. A patch antenna with band-notched characteristics at WLAN band for UWB applications is proposed in [31]. The band-notched characteristics at WLAN band are realized by proposing a square slot above the patch.

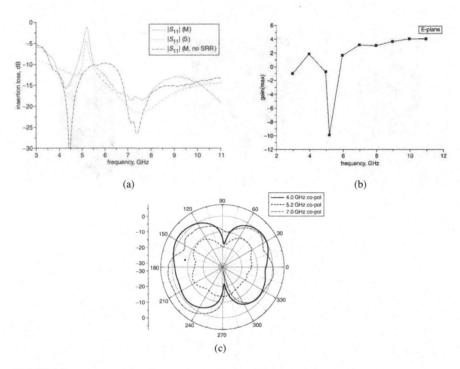

**FIGURE 2.2**
Simulated and measured results for slot loaded single band notched UWB antenna [19] (a) Simulated and measured return loss, (b) Gain versus frequency graph, and (c) Simulated co-pol. and cross pol. at different frequencies.

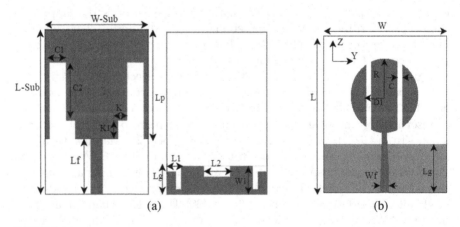

**FIGURE 2.3**
Different types of slot loaded single band notched UWB antennas [re-drawn] (a) Modified rectangular slots antenna [26], and (b) Segmented circular monopole antenna [27].

Jainal et al. [32] have designed a UWB planar antenna comprising slanting slot over a circular radiating patch for realizing notched band at WLAN band. Biswas et al. [33] have presented a fractal UWB antenna using third iterative modified Sierpinski carpet pattern slots on a circular monopole. Better performance with respect to the impedance bandwidth is achieved by truncating the ground plane in steps as well as by using fractal slots on the radiator. Further, incorporation of U-shaped slot within the ground plane makes the structure band-notched for the frequency range of 5–6 GHz. In [34], a UWB planar monopole antenna with band-notch properties is presented. The presented antenna comprises of a trapezoid shape with tapered end radiating patch. By introducing U-shaped slot, the notch characteristic is realized ranging from 5 GHz to 6 GHz. A UWB monopole microstrip patch antenna with enhanced bandwidth and narrow notched band characteristics is presented in [35]. Thereafter, introduction of two configurations of C-shaped slots on the partial ground plane causes the generation of desired single notched band characteristics (6.26–7.28 GHz). Likewise, by incorporating U-shaped slot on the radiating patch, notched band behavior within 3.1–3.2 GHz has been realized [36]. A recent design of balanced antipodal tapered slot antenna (BATSA) with notched band characteristics using quarter-wavelength spur line is presented in [37]. A notched band wearable UWB monopole antenna has been proposed in [38]. The notch band 6.3–7.15 GHz is realized by using split ring slots. Moreover, the presence of the full ground plane makes the proposed UWB antenna appropriate for wearing applications by decreasing specific absorption rate values. Finally, in Table 2.1, a comparison has been made with respect to existing slot-loaded UWB antennas.

**TABLE 2.1**

Slotted UWB Designs for Different Single-Notched Applications

| Ref. | Types of Slot | Notch-Band/ Notch-Frequency (GHz) | Notched-Band Application | Volume (mm$^3$) |
|---|---|---|---|---|
| [13] | Rectangular | 4.80–5.50 | WLAN | 26.5×21.5×1.6 (= 911) |
| [14] | Half wave resonant structure | 5.6 | WLAN | 34×27×1.6 (= 1468) |
| [15] | Split ring resonator (SRR) | 5.04–6.09 | WLAN | 42×46×1.6 (= 3091) |
| [16] | V-shaped | 5.15–5.35 | WLAN | 32×22×0.036 (= 25.34) |
| [19] | Split ring resonator (SRR) | 5.15–5.35 | WLAN | 25×25×0.762 (= 476) |
| [22] | Two slits | 5.0 | WLAN | 31×20×0.05 (= 31) |
| [25] | Shovel-shaped | 5.13–6.1 | WLAN | 18×15×1 (= 270) |
| [26] | Modified rectangular | 5.02–5.9 | WLAN | 20×12×1.6 (= 384) |
| [27] | Segmented resonator | (5.1–5.6) | WLAN | 47×37×1.5 (= 2608) |
| [28] | Arc-shaped | (4.9–5.9) | WLAN | 30×30×1 (= 900) |
| [30] | U-shaped | (5.1–5.8) | WLAN | 34×40×1.6 (= 2176) |
| [34] | U-shaped | (5–6) | WLAN | 24×26×1.6 (= 998) |
| [35] | Inverted dome | (6.26–7.28) | INSAT | 23×35×1.6 (= 1288) |

## 2.3  Parasitic-Stub Loaded Geometries

For developing band-notched property, one of the most popular approaches is to incorporate parasitic elements or stub-type elements near the conducting region. A new band-notched PMA is shown in Figure 2.4(a). The notched band characteristic for WLAN 5-GHz band is obtained by applying a compact CPW resonant cell (CCRC) [39]. Next, a simple and compact microstrip-fed UWB PMA with band-notch characteristics is proposed in [40]. The antenna is composed of two monopoles of the same size and a small strip bar, which leads to the desired high attenuation at the notch frequency. Yoon et al. [41] have proposed a novel UWB wide slot antenna that has both enhanced impedance bandwidth and band rejection characteristics at 5 GHz. To achieve these, open stub elements and tapered microstrip feed line and conductor lines in the slot are used. A new compact UWB patch antenna with band rejection function is presented in [42]. By introducing a slotted-parasitic

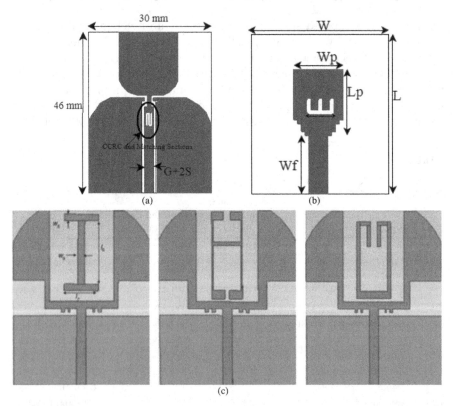

**FIGURE 2.4**
Different types of stub and parasitic element loaded single band notched antennas [re-drawn] (a) CCRC antenna [39], (b) Slotted-parasitic patch antenna [42], and (c) I/H/U shaped parasitic strip loaded antenna [45].

patch on the bottom layer (see Figure 2.4(b)) of the antenna, notched band characteristics have been achieved. A novel modified microstrip-fed UWB planar monopole antenna with variable frequency band-notch characteristics (5.1–5.6 GHz) has been designed by Zaker et al. [43]. Similarly, by using an inverted-cup strip, a notched band function at 5.0 GHz is accomplished in [44]. A compact printed UWB antenna with band-notched characteristics (5–6 GHz) is realized in [45] by introducing a resonant I-shaped strip, H-shaped strip, and U-shaped strip on the radiator, which is shown in Figure 2.4(c), and its performance parameters are shown in Figure 2.5. A circular ring antenna for UWB applications with band-notch performance is designed by Ghosh [46]. The band-notched characteristic is achieved by introducing a tuning stub inside the ring monopole.

Ojaroudi et al. [47] have proposed a novel printed monopole band-notched UWB antenna with a fractional bandwidth of more than 130%. The bandstop characteristic is obtained by using the square-ring radiating patch with a pair of T-shaped strips. A new UWB antenna with notched band characteristics (5.4–5.95 GHz) is designed using a rectangular stub [48]. A miniaturized UWB antenna with band rejection characteristics is presented in [49, 50]. The band rejection behavior within 5–6 GHz is realized by using arc-shaped

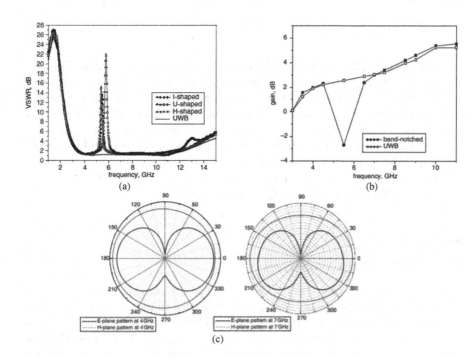

**FIGURE 2.5**
Simulated and measured results for parasitic strips loaded single band notch antenna [45] (a) VSWR curve of the UWB and band-notched antenna (b) Gain curve of the UWB and band-notched antenna, and (c) Radiation pattern at different frequencies.

**TABLE 2.2**

Parasitic stub/Strip-Based UWB Antenna Designs for Different Single-Notched Applications

| Ref. | Design Approach | Notch Band Frequency (GHz) | Notch Band Applications | Overall Size (mm³) |
|------|-----------------|----------------------------|-------------------------|--------------------|
| [39] | CCRC | (5.10–5.94) | WLAN | 30×40×1 (= 1200) |
| [40] | Small strip bar | (4.9–6.0) | C-band Downlink | NA |
| [41] | Open stub | (5.15–5.825) | WLAN | 30×26×0.5 (= 390) |
| [42] | Slotted parasitic | (5.15–5.825) | WLAN | 35×30×1.27 (= 1333) |
| [43] | H-shaped conductor | (5.1–5.9) | WLAN | 22×22×1 (= 484) |
| [44] | Inverted cup-striped | (5.15–5.825) | WLAN | 50×50×0.8 (= 2000) |
| [45] | I/H/U-shaped stripped | (5–6) | WLAN | 20×25×1 (= 500) |
| [46] | Tuning STUB | (4.0–4.6) | C-band Downlink | NA |
| [47] | T-shaped strips | (5.05–5.95) | WLAN | 12×18×1.6 (= 345) |
| [48] | Split ring parasitic element | (5.4–5.9) | WLAN | 30×30×1.6(= 1440) |
| [49] | Arc-shaped parasitic element | (5–6) | WLAN | 38×39×0.813(=1204) |
| [50] | S-shaped | (5–6) | WLAN | 20×20×1.5(= 600) |
| [51] | H-shaped strip | (5–5.6) | WLAN | 30×30×1(= 900) |

parasitic strip in [49] and S-shaped parasitic strip in [50]. Kavita et al. [51] have designed a band-notch UWB antenna using H-shaped strip that is placed on the back of a conventional UWB antenna. In Table 2.2, a comparison has been made of existing parasitic stubs-loaded UWB antennas. Despite all the advantages of afore-discussed techniques such as compact resonant structure and easy implementation, they suffer from poor notched width controlling capabilities, and poor radiation pattern due to irregularities in structure. To overcome the issues, EBG structures are being recently used for realizing band-notched characteristics in UWB antennas.

## 2.4 EBG-Loaded Geometries

The EBG structures are periodic collections of dielectric materials and conductors. The concept of EBG structures originates from the solid-state physics domain [57]. The bandgap feature of EBG has revealed the suppression of surface-wave in a particular frequency band, which results in improving the performance of the antenna. Another feature of the reflection phase of EBG structure can be intended to realize a perfect magnetic conductor like surface in a certain frequency band by using the reflection property. The reflection phase of an EBG surface varies continuously from +180° to –180° [58]. Due to

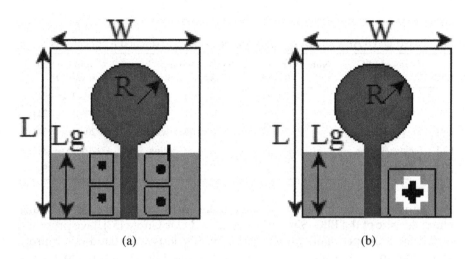

**FIGURE 2.6**
EBG loaded single-notch UWB antennas [re-drawn] (a) Mushroom type EBG [52], and (b) Plus shaped EBG structure [54].

its unique bandgap property, the EBG structures are widely used in several applications, viz. gain improvement [59], mutual coupling reduction [60], and simultaneous switching noise suppression [61]. Further, the EBG structures are also been used in UWB antennas to reject the interference that are discussed further. Yazdi and Komjani [52] have designed a new compact UWB circular monopole antenna with band rejection at 5.5 GHz. This rejection band is created by means of a mushroom-type EBG structure which is shown in Figure 2.6(a), and its performance parameters are shown in Figure 2.7.

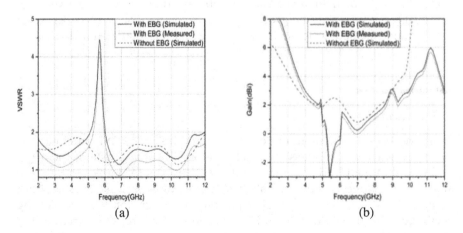

**FIGURE 2.7**
Simulated performance of mushroom type EBG loaded single-notch antenna [52] (a) VSWR performance, and (b) Gain performance.

**TABLE 2.3**

EBG-Based UWB Antenna Designs for Different Single-Notched Applications

| Ref. | Types of EBG Structure | Notch Band Frequency (GHz) | Notch Band Applications | Geometry Size (mm³) |
|------|------------------------|----------------------------|-------------------------|---------------------|
| [52] | Mushroom-type | (5.29–5.65) | WLAN | 39×35×0.3125 (= 426) |
| [53] | Mushroom-type | (5–6) | WLAN | 42×20×1.6 (= 1344) |
| [54] | Plus-shaped | (7.25–7.75) | X-band downlink | 42×20×1.6 (= 1344) |
| [55] | Square/Circle | (6.05–7.43), (6.23–7.68) | C-band Uplink | 42×20×1.6 (= 1344) |
| [56] | C-shaped | (6.21–7.2) | C-band Uplink | 40×38.5×1 (= 1540) |

An equivalent circuit model is also employed to investigate the stopband characteristic of the EBG. Similarly, Jaglan and Dev Gupta [53] have proposed a circular monopole antenna for UWB applications with band-notch property. Antenna utilizes modified mushroom-type EBG structures to achieve band-notched at WLAN 5–6 GHz band.

A UWB antenna with band notch property is realized by means of uniplanar EBG structure (See Figure 2.6(b)) that is capacitively attached to antenna [54]. Antenna integrated with plus-shaped EBG discards the frequency band 7.25–7.75 GHz used for satellite downlink signal. Moreover, notch occurrences may be altered by varying the attaching outlines and cavities of EBG cells. In [55], a UWB circular monopole antenna is designed using uni-planar EBG structures. Two different configurations of EBG structures have been presented for realizing notched within the UWB region. Antenna with square and circular EBGs has a notch in C-band at frequencies 6.05–7.43 GHz and 6.23–7.68 GHz, respectively. Song et al. [56] have designed a new band-notched UWB antenna based on EBG technology. The EBG structure is applied in a UWB antenna to create a stopband, and it is located between the microstrip feedline and the ground plane. A comparison of the EBG based single band-notched designs has been made and it is tabulated in Table 2.3.

## 2.5 Summary

In above sections, the literature on single band-notched UWB antennas has been presented and discussed in detail. From the above state of the art, it is found that the UWB antennas need to be compact for implementation in the modern wireless communication system. However, it is also necessary to have UWB antenna with high rejection characteristics. Various slotting techniques have been discussed here for single-band rejection characteristics. Furthermore, band-notched behavior based on EBG structures is also discussed, but the literature is very limited.

# References

[1] FCC, Washington, DC, Federal Communications commission revision of part 15 of the commission's rules regarding ultra-wideband transmission systems. First reported Order FCC: 02. V48, 2002.

[2] A. M. Abbosh, "Design of a CPW-fed band-notched UWB antenna using a feeder-embedded slot-line resonator," International Journal of Antennas and Propagation, vol. 2008, pp. 1–5, 2008.

[3] J. Liang, Antenna Study and Design for Ultra Wideband Communication Applications, United Kingdom: University of London, 2006.

[4] H. G. Schantz, "Introduction to Ultra-Wide Band Antennas," In: Proceedings from IEEE Conference Ultra Wideband System Technologies, Reston, USA, pp. 1–9, November, 2003.

[5] C. A. Balanis, Antenna Theory Analysis and Design, New York: John Wiley & Sons, 3rd ed., chap. 14, 2009.

[6] R. Garg, P. Bhartia, I. Bahl, A. Ittipiboon, Microstrip Antenna Design Handbook, Norwood, Massachusetts: Artech House, 2001.

[7] A. Ghosh, M. Gangopadhyay, "Bandwidth optimization of microstrip patch antenna- A basic overview," International Journal on Recent and Innovation Trends in Computing and Communication, vol. 6, no. 2, p. 133, 2018.

[8] F. Mouhouche, A. Azrar, M. Dehmas, K. Djafri, "Compact Dual-Band Reject UWB Monopole Antenna Using EBG Structures," In: International Conference on Electrical Engineering-Boumerdes (ICEE-B), Boumerdes, pp. 1–5, 2017.

[9] X. Hu, X. Yang, "Tri-Band-Notched Ultrawideband (UWB) Antenna Using Split Semicircular Resonator (SSR) and Capacitively Loaded loOps (CLL)," In: IEEE International Conference on Communication Problem-Solving (ICCP), Guilin, pp. 186–189, 2015.

[10] D. Rosepriya, Kayalvizhi, "Compact Design of Slot Antenna with Band-Notched Function for UWB Application," In: International Conference on Electronics and Communication Systems (ICECS), Coimbatore, pp. 1012–1016, 2015.

[11] K. Yu, Y. Li, X. Luo, X. Liu, "A planar UWB antenna with quad notched bands using rake-shaped resonator and L-shaped slots," IEEE International Symposium on Antennas and Propagation (APSURSI), Fajardo, pp. 1815–1816, 2016.

[12] T. Dissanayake, K. P. Esselle, "Design of slot loaded band-notched UWB antennas," IEEE Antennas and Propagation Society International Symposium, Washington, DC, pp. 545–548, vol. 1B, 2005.

[13] P. Rakluea, J. Nakasuwan, "Planar UWB antenna with single band-notched characteristic," ICCAS 2010, Gyeonggi, pp. 1978–1981, 2010.

[14] S. Jacob, P. Mohanan, "UWB antenna with single notch-band for WLAN environment," Indian Antenna Week (IAW), Kolkata, pp. 1–4, 2011.

[15] J. K. Deegwal, A. Kumar, S. Yadav, M. M. Sharma, M. C. Govil, "Ultra-wideband truncated rectangular monopole antenna with band-notched characteristics," IEEE Symposium on Wireless Technology and Applications (ISWTA), Bandung, pp. 254–257, 2012.

[16] Y. Kim, D. Kwon, "CPW-fed planar ultra wideband antenna having a frequency band notch function," Electronics Letters, vol. 40, no. 7, pp. 403–405, 2004.

[17] S. Nikolaou, B. Kim, Y. Kim, J. Papapolymerou, M. M. Tentzeris, "CPW-fed Ultra Wideband (UWB) Monopoles with Band Rejection Characteristic on Ultra-Thin Organic Substrate," In: Asia-Pacific Microwave Conference, Yokohama, pp. 2010–2013, 2006.

[18] Y. Kim, D. H. Kwon, "Planar ultra wideband slot antenna with frequency band notch function," IEEE Antennas and Propagation Society Symposium, Monterey, CA, USA, vol. 2, pp. 1788–1791, 2004.

[19] J. Kim, C. S. Cho, J. W. Lee, "5.2 GHz notched ultra-wideband antenna using slot-type SRR," Electronics Letters, vol. 42, no. 6, pp. 315–316, 2006.

[20] Q. Chu, Y. Yang, "A Compact CPW-fed Planar Ultra-Wideband Antenna with a Frequency Notch Characteristic," In: Asia-Pacific Microwave Conference, Bangkok, pp. 1–4, 2007.

[21] A. H. M. Z. Alam, M. R. Islam, S. Khan, "Designing an UWB Patch Antenna with Band Notched by Using L-shaped Slot and Unsymmetrical Feedline," In: Canadian Conference on Electrical and Computer Engineering, Niagara Falls, ON, pp. 101–104, 2008.

[22] S. W. Bae, H. K. Yoon, W. S. Kang, Y. J. Yoon, C. Lee, "A Flexible Monopole Antenna with Band-Notch Function for UWB Systems," In: Asia-Pacific Microwave Conference, Bangkok, pp. 1–4, 2007.

[23] Z. Li, Y. Wang, J. Wang, K. Jiang, "An Ultra-Wideband Planar Monopole Antenna with Band-Notched Characteristics," In: International Conference on Microwave Technology and Computational Electromagnetics (ICMTCE 2009), Beijing, pp. 62–65, 2009.

[24] M. M. Sharma, A. Kumar, S. Yadav, Y. Ranga, D. Bhatnagar, "A compact ultra-wideband CPW-fed printed antenna with SRR for rejecting WLAN band," Indian Antenna Week (IAW), Kolkata, pp. 1–3, 2011.

[25] A. Nouri, G. R. Dadashzadeh, "A compact UWB band-notched printed monopole antenna with defected Ground structure," IEEE Antennas and Wireless Propagation Letters, vol. 10, pp. 1178–1181, 2011.

[26] R. Movahedinia, M. Ojaroudi, S. S. Madani, "Small modified monopole antenna for ultra-wideband application with desired frequency band-notch function," IET Microwaves, Antennas & Propagation, vol. 5, no. 11, pp. 1380–1385, 2011.

[27] K. Zhang, Y. Li, Y. Long, "Band-notched UWB printed monopole antenna with a novel segmented circular patch," IEEE Antenna Wireless Propagation Letters, vol. 9, pp. 1209–1212, 2010.

[28] Y. Jiang, H. Zhang, H. Xu, R. Zhang, X. Zeng, "A Novel Ultra-Wideband Antenna with Band Notch Characteristic," In: International Conference on Microwave and Millimeter Wave Technology (ICMMT), Shenzhen, pp. 1–4, 2012.

[29] F. E. Mahmood, H. M. Alsabbagh, R. Edwards, "CPW-Fed UWB Antenna with Band-Notch by Hexagonal Shape Slot," In: International Conference on Future Communication Networks, Baghdad, pp. 69–71, 2012.

[30] Y. Hacene, X. Shuguo, T. Rahman, "Design of a novel monopole antenna with 5.5 GHz band-notch characterization for UWB applications," Xian, pp. 273–276, 2012.

[31] S. Jangid, M. Kumar, "A Novel UWB Band Notched Rectangular Patch Antenna with Square Slot," In: International Conference on Computational Intelligence and Communication Networks, Mathura, pp. 5–9, 2012.

[32] S. F. Jainal, T. Wakabayashi, O. Ayob, M. K. A. Rahim, "A UWB Planar Antenna Comprising a Single Slot Elliptical Element with Band Notch Characteristics," In: IEEE International RF and Microwave Conference (RFM), Penang, pp. 133–137, 2013.

[33] B. Biswas, D. R. Poddar, R. Ghatak, A. Karmakar, "Modified Sierpinski Carpet Fractal Shaped Slotted UWB Monopole Antenna with Band Notch Characteristic," In: National Conference on Communications (NCC), New Delhi, India, pp. 1–5, 2013.

[34] M. A. Matin, M. M. Hossain, "A New Planar Printed Antenna with Band-notch Characteristics for UWB Applications," In: IEEE Conference, TENCON, pp. 1–3.

[35] M. A. S. Al-Zahrani, O. I. S. Al-Qahtani, F. D. M. Al-Sheheri, A. S. M. Qarhosh, A. M. Al-Zahrani, M. S. Soliman, "Design and performance analysis of an ultra-wideband monopole microstrip patch antenna with enhanced bandwidth and single band-notched characteristics," Progress in Electromagnetics Research Symposium, Singapore, pp. 1488–1493, 2017.

[36] K. J. Singh, R. Mishra, "Design of A Circular Microstrip Patch Antenna with Single Band Notch Characteristic for UWB Applications," In: IEEE International WIE Conference on Electrical and Computer Engineering (WIECON-ECE), Dehradun, pp. 262–265, 2017.

[37] C. Sarkar, C. Saha, L. A. Shaik, J. Y. Siddiqui,Y. M. M. Antar, "Frequency notched balanced antipodal tapered slot antenna with very low cross-polarized radiation," IET Microwaves, Antennas & Propagation, vol. 12, no. 11, pp. 1859–1863, 2018.

[38] P. Rahmatian, E. Moradi, M. Movahhedi, "Single Notch Band UWB Off-Body Wearable Antenna with Full Ground Plane," In: Iranian Conference on Electrical Engineering (ICEE), Yazd, Iran, pp. 1228–1232, 2019.

[39] S. Qu, J. Li, Q. Xue, "A band-notched ultra wideband printed monopole antenna," IEEE Antennas and Wireless Propagation Letters, vol. 5, pp. 495–498, 2006.

[40] S. Hong, K. Chung, J. Choi, "Design of a band-notched wideband antenna for UWB applications," IEEE Antennas and Propagation Society International Symposium, Albuquerque, NM, pp. 1709–1712, 2006.

[41] H. K. Yoon, Y. Lim, W. Lee, Y. J. Yoon, S. M. Han, Y. H. Kim, "UWB wide slot antenna with band-notch function," IEEE Antennas and Propagation Society International Symposium, Albuquerque, NM, pp. 3059–3062, 2006.

[42] N. H. M. Sobli, H. E. Abd-El-Raouf, "Design of a compact band-notched antenna for ultra wideband communication," IEEE Antennas and Propagation Society International Symposium, San Diego, CA, pp. 1–4, 2008.

[43] R. Zaker, C. Ghobadi, J. Nourinia, "Novel modified UWB planar monopole antenna with variable frequency band-notch function," IEEE Antennas and Wireless Propagation Letters, vol. 7, pp. 112–114, 2008.

[44] A. Kalteh, R. Fallahi, M. G. Roozbahani, "A novel microstrip-fed UWB circular slot antenna with 5-GHz band-notch characteristics," In: IEEE International Conference on Ultra-Wideband, Hannover, pp. 117–120, 2008.

[45] L. Y. Cai, G. Zeng, H. C. Yang, X. W. Zhan, "Compact printed ultra-wideband antennas with band-notched characteristics," Electronics Letters, vol. 46, no. 12, pp. 817–819, 2010.

[46] S. Ghosh, "Band-notched modified circular ring monopole antenna for ultra wideband applications," IEEE Antennas and Wireless Propagation Letters, vol. 9, pp. 276–279, 2010.

[47] M. Ojaroudi, S. Yazdanifard, N. Ojaroudi, R. A. Sadeghzadeh, "Band-notched small square-ring antenna with a pair of T-shaped strips protruded inside the square ring for UWB applications," IEEE Antennas and Wireless Propagation Letters, vol. 10, pp. 227–230, 2011.

[48] E. E. M. Khaled, A. A. R. Saad, D. A. Salem, "A Proximity-fed Ultra-Wideband Annular Slot Antenna with Band-Notch Characteristics via a Split-Ring Parasitic Element," In: European Conference on Antennas and Propagation (EUCAP), Prague, pp. 1–5, 2012.

[49] F. Zhu et al., "Miniaturized tapered slot antenna with Signal rejection in 5-6 GHz band using a balun," IEEE Antennas and Wireless Propagation Letters, vol. 11, pp. 507–510, 2012.

[50] M. Ojaroudi, N. Ojaroudi, S. Amiri, F. Geran, "Band-notched small microstrip slot antenna by using parasitic structures inside the slots for UWB applications," Telecommunications Forum (TELFOR), Belgrade, pp. 1168–1170, 2012.

[51] K. Ruchandani, M. Kumar, "A Novel CPW Fed Octagonal Aperture UWB Antenna with 5GHz/6GHz Band Rejection by H Shaped Slot," In: International Conference on Computer, Commun and Control (IC4), Indore, pp. 1–4, 2015.

[52] M. Yazdi, N. Komjani, "Design of a band-notched UWB monopole antenna by means of an EBG structure," IEEE Antennas and Wireless Propagation Letters, vol. 10, pp. 170–173, 2011.

[53] N. Jaglan, S. D. Gupta, "Reflection phase characteristics of EBG structures and WLAN band notched circular monopole antenna design," International Journal on Communications Antenna and Propagation (IRECAP), vol. 5, pp. 233–240, 2015.

[54] N. Jaglan, S. D. Gupta, S. Srivastava, B. K. Kanaujia, "Satellite Downlink Communication Band Notched UWB Antenna Using Uniplanar EBG Structure," In: International Conference on Signal Processing and Communication (ICSC), Noida, pp. 89–94, 2016.

[55] N. Jaglan, S. D. Gupta, S. Srivastava, " Notched UWB circular monopole antenna design using uni-planar EBG structures," International Journal on Communications Antenna and Propagation (IRECAP), vol. 6, no. 5, p. 266. doi:10.15866/irecap.v6i5.9456.

[56] C. Y. Song, T. Y. Yang, W. W. Lin, X. L. Yang, "Design of a Band-Notched UWB Antenna Based on EBG Structure," In: IEEE International Conference on Applied Superconductivity and Electromagnetic Devices, Beijing, pp. 274–277, 2013.

[57] F. Yang, Y. R. Samii, Electromagnetic Band Gap Structures in Antenna Engineering. New York: Cambridge University Press, 2009.

[58] S. Raza, "Characterization of the reflection and dispersion properties of mushroom-related structures and their application to antennas," Thesis on Master of Applied Science, University of Toronto, 2012.

[59] P. Ketkuntod, T. Hongnara, W. Thaiwirot, P. Akkaraekthalin, "Gain enhancement of microstrip patch antenna using I-shaped mushroom-like EBG structure for WLAN application," International Symposium on Antennas and Propagation (ISAP), Phuket, pp. 1–2, 2017.

[60] H. Sajjad, S. Khan, E. Arvas, "Mutual coupling reduction in array elements using EBG structures," International Applied Computational Electromagnetics Society Symposium-Italy (ACES), pp. 1–2, 2017.

[61] L. Shi, K. Li, H. Hu, S. Chen, "Novel L-EBG embedded structure for the suppression of SSN," IEEE Transactions on Electromagnetic Compatibility, vol. 58, no. 1, pp. 241–248, 2016.

# 3

## Dual Band-Notched UWB Antennas

### 3.1 Introduction

The US-Federal Communications Commission released 3.1–10.6 GHz unlicensed band for commercial UWB applications in 2002. Since then, the UWB technology has become very much popular for short-range wireless services [1, 2]. Due to their splendid features such as wide bandwidth, low power consumption, high data rates, easy fabrication, and others, printed UWB antennas are considered as promising elements for front-end devices in UWB communication systems [3]. However, in practical applications, the issue of bandwidth overlapping with the existing wireless technologies takes place. Some narrowband services such as WiMAX band (3.3–3.7 GHz), WLAN band (5.15–5.35 GHz and 5.725–5.825 GHz), C-band satellite communications bands (3.77–4.2 GHz for downlink and 5.9–6.4 GHz for uplink), and X-band ITU band (8.025–8.4 GHz) already occupy frequencies within the UWB frequency range [4–8]. These interfering bands may cause serious degradation in antenna performance. So, to reject the interfering bands, it is very much necessary to design an antenna with multi-band filtering capabilities. Various design techniques have been introduced to obtain band-notched characteristics that are discussed in the following section.

The work presents a detailed study on dual notched-band characteristics in UWB antennas. In general, a single parasitic element or slot can generate only one notched band and fails to reject multiple interferences. In order to realize a dual band-notched UWB antenna, techniques employing multiple resonant elements such as slotted geometries [9–47], parasitic elements [48–58], and EBG-based techniques [59–69] are commonly used.

### 3.2 Slots-Loaded Geometries

The band-notch characteristics in UWB antennas can be accomplished by loading slots in the radiating patch as well as in the ground plane. The slots in the radiating patch and ground plane cause disturbance in the flow of current, which leads to a reduction in antenna size and also gives rise to multiband operation. Similarly, it can also be utilized for realizing the notch frequency function in the UWB antennas. Various types of slots such as U-slot, L-slot, ring slot, etc., are being used for filtering out the narrowband frequencies. In [9], a compact microstrip-fed planar UWB antenna with dual band-notch characteristics is presented. By inserting an E-shaped slot in the radiating patch and a U-slot in the ground plane, dual notched-band characteristics at 3.49–4.12 GHz, and 5.66–6.43 GHz have been obtained. Moreover, the two notched-bands can be controlled by adjusting the length of the corresponding slots. Similarly, by using two nested C-shaped slots in a beveled rectangular patch, dual band rejections at 3.4 GHz and 5.5 GHz have been realized [10]. The equivalent circuit model of the proposed antenna is also presented to discuss about the mechanism of band-notch creation. A UWB antenna fed by microstrip-line with dual band-notched characteristics is presented in [11]. By inserting a C-shaped slot on the radiating patch and embedding a pair of inverted L-shaped slots into the ground plane, dual band-notched characteristics at 3.5–5.5 GHz are accomplished which is shown in Figure 3.1(a). Next, a dual band-notched monopole antenna with a circular disc-shaped patch is

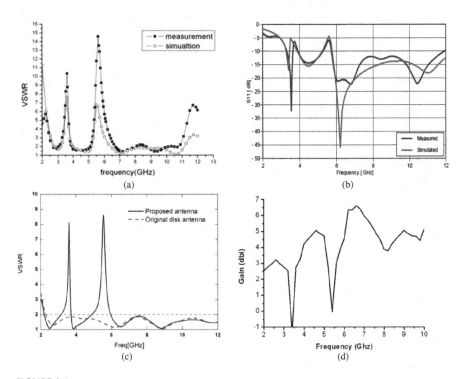

**FIGURE 3.1**
Simulated and measured results for slot loaded dual band notched UWB antennas (a) Simulated and measured VSWR curve [11], (b) Simulated and measured return loss [17], (c) Simulated VSWR curve [18], and (d) Gain versus frequency graph [18].

studied in [12]. The antenna consists of a circular patch with a C-shaped slot and a U-shaped slot on the front side of the substrate, which leads to the generation of dual notched-band frequencies at 3.5 GHz and 5.5 GHz, respectively. Li et al. [13] have proposed a CPW-fed UWB antenna with dual band-notch characteristics. The antenna is compact and has an overall size of 21×28×1.6 mm$^3$. By cutting two U-shaped slots in the radiation patch and H-shaped slot in the CPW ground, two band-notched frequencies at 5–6 GHz for WLAN and at 7.7–8.5 GHz for X-band are obtained. A novel UWB antenna with L-shaped slots on the substrates is presented in [14]. The proposed antenna produces an impedance bandwidth of 2.6–12 GHz with notch bands of 3.65 GHz and 5.5 GHz and has omnidirectional radiation pattern in the H-plane.

A compact planar monopole antenna with dual band-notched characteristics at 3.2–3.6 GHz and 5.05–6.08 GHz is accomplished by implementing two partial and semi-circular annular slots [15]. Furthermore, the suggested antenna has maintained the wideband performance from 3 GHz to 11 GHz. A planar monopole UWB antenna with dual band-notched characteristics is presented in [16]. The notched-band characteristics at 5.3 GHz and 7.4 GHz are achieved by etching one quasi-complimentary split ring resonator (CSRR) in the feed line. Tammam et al. [17] have designed a low-profile monopole UWB antenna with band-rejection characteristics. By utilizing a half wavelength slot on the radiator, the suggested antenna rejects frequency in the WiMAX band. Further, additional notch has been realized in the WLAN band by using two symmetrical open-ended quarter-wavelength slots in the ground plane. The performance parameters for [17] are shown in Figure 3.1(b). Zhao et al. [18] have presented a novel monopole antenna with dual band-notched behavior. To obtain this feature, two optimized arc-shaped slots are embedded in the disk patch and its simulated and measured performance parameters are shown in Figure 3.1(c) and (d). A novel printed monopole antenna for UWB applications with dual band-notch function is presented in [19]. The antenna consists of a square radiating patch with a pair of L-shaped slits and an E-shaped slot, shown in Figure 3.2(a), to produce dual band-notch function at 3.5 and 5.5 GHz, respectively. Moreover, a V-shaped protruded strip is used in the ground plane which provides wide impedance bandwidth of more than 140% (2.89–17.83 GHz). Similarly, by introducing U- and C-shaped slots on the radiation patch, dual band rejection property in the 3.5–5.5 GHz band is obtained in [20].

A dual band-notched UWB slot antenna, fed by CPW technique, is presented in [21]. The notched bands at 5.8 GHz WLAN and 8 GHz X-band frequencies are accomplished by inserting a pair of slots in the radiation patch. Ojaroudi et al. [22] have proposed a novel method for designing a UWB monopole antenna with dual band-notched characteristics. The proposed antenna consists of a square radiating patch with a modified T-shaped slot and a ground plane with two E-shaped slots and a W-shaped conductor backed-plane. The modified T-shaped slot over the radiating patch is responsible for generating dual notched-band characteristics for the applications of WiMAX and WLAN bands. Likewise, by using SRRs, two notches at frequencies 3.5 GHz

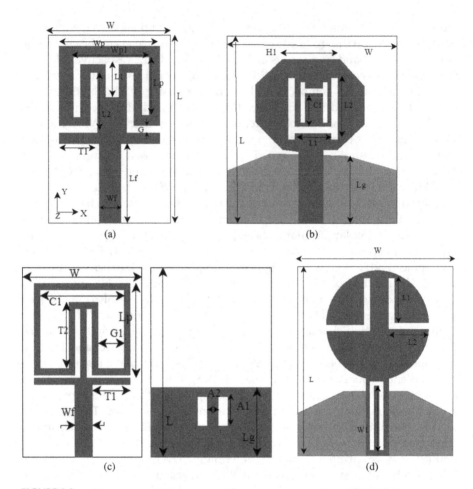

**FIGURE 3.2**
Different types of slot loaded dual band notched UWB antennas [re-Drawn] (a) L and E-slots antenna [19], (b) H-shaped slot antenna [25], (c) U-ring strip antenna [26], and (d) Elliptical monopole antenna [28].

and 5.3 GHz are obtained by Kulkarni et al. [23]. A new planar antenna for UWB-applications with two band notches behavior is presented in [24]. The antenna is designed to have rejection over a frequency span of 3.3–3.8 GHz and 5.15–5.825 GHz by connecting T-shaped strip to the top of the patch and a complementary circular slot-type SRR at the bottom of the patch. Similarly, by inserting two altered H-shaped slots on the octagon radiating patch (See Figure 3.2(b)), dual band-notches at 3.3–3.6 GHz and 5–5.8 GHz are obtained [25].

Figure 3.2(c) reveals a novel, small, printed monopole antenna for UWB applications with dual band-notch functions using a coupled inverted U-ring strip in the radiating patch. Besides this, the antenna has achieved a size reduction of

28% [26]. A new UWB antenna using an inverted F-slot in the patch and a U-slot in the microstrip feed line is designed [27]. The suggested antenna covers an impedance bandwidth of 3.1–10.6 GHz and has the capability to reduce interference of 5.15–5.825 GHz and 7.2–8.4 GHz bands, respectively. Mohammed et al. [28] designed a UWB printed elliptical monopole antenna with dual notched-band functions. The notched-band characteristics are achieved by utilizing symmetrical L-shaped slot and an inverted U-shaped slot in the radiating patch and feed line, which is shown in Figure 3.2(d). A novel heart-shaped dual band-notched UWB antenna is presented in [29]. The antenna consists of a heart-shaped patch fed with a microstrip line to achieve UWB characteristics. The band-notched characteristic is achieved by using a U-shaped slot on the patch and a pair of C-shaped stubs near the feeding line. Similarly, by using a J-shaped slot on the modified ground plane and a U-shaped slot on the feed line of a fork-shaped UWB antenna, dual-notched-band characteristics for WiMAX (3.4–3.69 GHz) and WLAN (5.15–5.85 GHz) bands have been realized [30]. A slot-shaped UWB monopole antenna with frequency rejections in WLAN and WiMAX bands is presented in [31]. The notched-band characteristics within 3.3–3.8 GHz (WiMAX) and 4.8–5.85 GHz (WLAN) are achieved by using two arc-shaped resonators and a cross-shaped resonator in the circular-ring-shaped monopole patch. Likewise, by using L- and U-shaped slot, dual band-notched characteristics for WLAN and C-band satellite communication systems have been realized by Yadav et al. [32]. A compact coplanar waveguide-fed ultra wideband antenna having frequency notch characteristics at 5.2 GHz for WLAN and 8.2 GHz for X-band applications has been proposed by using fractal-type hexagonal star slot [33]. In [34], a UWB antenna is proposed with dual band-notched characteristics at 4.7 GHz and 6.95 GHz. The filtering property is accomplished by modifying the circular patch into fractal types. A novel printed monopole antenna for UWB application with dual band-notch characteristics is presented in [35]. The proposed antenna consists of a small square radiating patch with $\pi$-shaped and W-shaped slots, which are responsible for avoiding interference at 3.3–4.2 GHz and 5.5–5.96 GHz within the UWB region, and a pair of C-shaped slots over the ground plane that provides wide impedance bandwidth. A dual-notch novel UWB monopole antenna is proposed in [36]. The suggested antenna gives a good impedance matching from 3.2 GHz to 40 GHz with dual band-rejection characteristics operating from 4.9 GHz to 6 GHz and 7.14 GHz to 8.42 GHz, respectively. Dual notch-band UWB printed antenna fed by CPW has been proposed in [37]. The notched-band characteristics at 3.7 GHz and 5.5 GHz are achieved by changing the structure with horn-shaped structure, equilateral triangle, and rectangular slot in the patch. In [38], a UWB antenna with dual band-notched characteristics at WiMAX and WLAN band is proposed. The suggested antenna has a modified rectangular radiating patch fed with a microstrip feedline. The band-stop filtering at WiMAX and WLAN is achieved by etching a rectangular slot in the defected ground structure (DGS) and a C-shaped slot in the modified metal patch. Guan et al. [39] have designed a novel UWB wearable antenna with dual notched-band property. Dual notched-band characteristics for WiMAX and WLAN applications are obtained by implementing defected microstrip (DMS) slot on the microstrip feed line, and DGS down to the feed line on the ground plane. An elliptical UWB patch antenna with partial arc-shaped ground plane fed by micro-strip transmission line is proposed and depicted in Figure 3.3. Two open-ended quarter-wavelength straight slots are etched in the patch to generate dual notched-band properties for the applications of WiMAX (3.1–4.8 GHz) and X-band (9.6–11.2 GHz), respectively. Manshouri et al. [41] have designed a microstrip-fed UWB antenna with a simple architecture that operates at a frequency range of 3.1–11 GHz. By embedding an SRR on the patch, a stopband between 4.8 GHz and 5.9 GHz has been successfully realized. Furthermore, an additional notch at 7.5–8.5 GHz has been obtained by using a CSRR.

A UWB antenna with dual-band rejection characteristics and size miniaturization approach has been presented in [42]. The proposed antenna consists of a unique patch of trapezoidal shape along with V-shaped slot to reject the interference of (3.15–4 GHz) and partial ground plane. Furthermore, a pair of C-shaped strips over the feed line are introduced to obtain notched-band

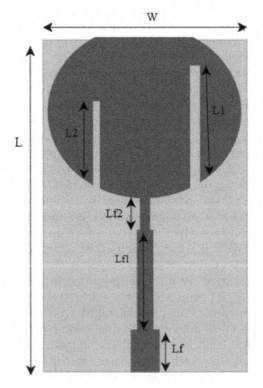

**FIGURE 3.3**
Elliptical patch antenna [40] [re-drawn].

characteristics over the WLAN (5.18–5.95 GHz) band. A UWB antenna with compact-circular monopole patch and semi-rectangular ground plane is presented in [43]. The proposed antenna is incorporated with L-shaped complimentary slots and I-shaped slots to produce notches for the applications of WiMAX and WLAN. A compact wearable textile antenna with dual band-notched characteristics for UWB applications is presented in [44]. The antenna consists of a CPW-fed semi-circular radiator with two semi-circular SRRs on the patch to achieve a notched band from 2.3 GHz to 2.5 GHz for Bluetooth application band, and the other notch band is obtained from 3.3 GHz to 3.7 GHz for WiMAX applications.

Soliman et al. [45] have designed a compact UWB patch antenna with dual narrow-band notched performance for stopping the activities of WiMAX band (3.3–3.7 GHz) and WLAN band (5.15–5.825 GHz) services by inserting a mirror L-shaped configuration and SRR on the patch. Kundu in [46] has designed a compact printed UWB antenna having impedance bandwidth from 2.78 GHz to 16.58 GHz with dual frequency stop bands at 3.3–3.81 GHz and 5.05–5.52 GHz to eliminate WiMAX and lower WLAN interferences by utilizing rectangular split ring-shaped slots on the circular patch. Similarly, by using via shorted CSRR metamaterial near the feedline dual band notched antenna is realized in [47]. The proposed antenna rejects dual interfering bands ranging from 5.1 GHz to 5.8 GHz and 3.3 GHz to 3.7 GHz, respectively. Finally, a comparison has been made in Table 3.1 with respect to slots-loaded dual band-notched UWB antennas.

**TABLE 3.1**

Slots-Loaded Dual Band-Notched UWB Antennas: A Comparison

| Ref. | Design Approach | Notch Frequency/ Notch-Band (GHz) | Notch-Band Applications | Overall Geometry (mm³) |
|---|---|---|---|---|
| [10] | C-shaped slots | 3.4, 5.5 | WiMAX, WLAN | 30×26×1.6 (= 1248) |
| [12] | U/C-shaped slots | 3.5, 5.5 | WiMAX, WLAN | 30×35×1.6 (= 1680) |
| [13] | U/H-shaped slots | 5–6, 7.7–8.5 | WLAN, X-band | 21×28×1.6 (= 940) |
| [16] | CSRR | 5.3, 7.4 | WLAN, X-band | 46.4×38.5×1 (= 1786) |
| [20] | U/C-shaped slots | 3.5, 5.5 | WiMAX, WLAN | 24×24×1 (= 576) |
| [21] | C-shaped slots | 5.8, 8.0 | WLAN, X-band | 25×24×0.787 (= 472) |
| [22] | T-shaped slots | 3.5–5.5, 5.2–5.8 | C-band, WLAN | 12×18×1.6 (= 345) |
| [23] | SRR | 3.5, 5.5 | WiMAX, WLAN | 36×40×1.6 (= 2304) |
| [24] | SRR | 3.3–3.8, 5.15–5.825 | WiMAX, WLAN | 30×30×1.6 (= 1440) |
| [26] | Inverted U-ring strips | 3.5, 5.5 | WiMAX, WLAN | 18×12×0.8 (= 172) |
| [28] | L-shaped slots | 3.4, 5.6 | WiMAX, WLAN | 22×25.5×1.6 (= 897) |
| [30] | J/U-shaped slots | 3.4–3.69, 5.15–5.85 | WiMAX, WLAN | 25×30×1.6 (= 1200) |
| [32] | U/L-shaped slots | 3.8–4.2, 5.1–5.8 | C-band, WLAN | 26×30×1.6 (= 1248) |
| [39] | DMS | 3.3–3.8, 5.15–5.85 | WiMAX, WLAN | 26×18×0.5 (= 234) |
| [41] | C/M-slots | 4.8–5.9, 7.5–8.5 | WLAN, X-band | 20×26×1.5 (= 780) |
| [44] | Semi-circular SRR | 2.3–2.5, 3.3–3.7 | Bluetooth, WiMAX | 43×40×2 (= 3440) |

## 3.3  Parasitic-Stub Based Geometries

Wide impedance bandwidth with dual notch-filtering characteristics in UWB antennas is a challenging task. The easiest way to accomplish this is by adding stubs and parasitic elements to the radiating patch and ground plane. Works proposed in [48–58] have implemented various stubs and parasitic elements to generate notches in UWB antennas. In [48], a novel UWB antenna with two controllable notch bands is proposed. The proposed antenna is fed by microstrip transmission line and provides wide bandwidth from 2.2 GHz to 12 GHz. By etching a split ring slot in the radiation patch, a stopband of 3.2–3.7 GHz has been achieved. Additional notched band from 5–6 GHz has also been realized by using a stepped impedance resonator (SIR). A planar slot antenna with dual band-notch characteristics is proposed for UWB application in [49]. The basic antenna comprises a rectangular tuning stub and a ground plane with a tapered slot. Therefore, to create notched-band characteristics for WiMAX applications, one angle-shaped parasitic slit is etched along with the tuning stub. Similarly, by placing two symmetrical parasitic slits inside the slot of the ground plane, the antenna produces a notched band for applications in WLAN bands.

Azim et al. [50] have designed a compact printed planar UWB antenna with dual band-notched properties. The proposed antenna consists of a rectangular radiating patch and a modified partial ground plane. Dual band-notched characteristics at 3.3–3.8 GHz and 5.1–5.6 GHz have been realized by placing two rectangular parasitic strips below the substrate which can be graphically analyzed from Figure 3.4(a) and (b). Moreover, stable radiation pattern and flat gain except in the notched bands have also been achieved. Azim et al. [51] have designed a compact V-shaped UWB antenna with dual band-notched characteristics. The dual band-notch functions at WiMAX band (3.17–3.85 GHz) and ITU band (7.9–9.1 GHz) are achieved using two parasitic stubs and an inverted U-shaped slot that are attached to the patch and the feedline as presented in Figure 3.5(a). Azim et al. [52] have also designed a dual notched-band (3.5 GHz and 5.5 GHz) UWB antenna. The dual band-notch function is achieved by etching a single tri-arm resonator below the patch (see Figure 3.4(c) and (d)). Similarly, by using inverted U- and iron-shaped parasitic resonators on the back side of the radiating patch (Figure 3.5(b)), dual-notched-band characteristic at 5.2 and 8.4 GHz have been obtained [53]. In [54], a planar UWB antenna with dual band-notched characteristic is proposed. Two I-shaped and one inverse E-shaped metal stubs are embedded into the broad slot, making band-notched characteristics in the frequency range between 3–3.9 GHz and 5–6 GHz, respectively. A microstrip-fed UWB planar monopole antenna with dual band-notched performance is presented in [55]. Two notched bands, covering 5–6 GHz WLAN band and 3.3–3.8 GHz WiMAX, are provided by using patches placed in the back of the substrate and connected to the main radiating patch by a series of via holes. Besides this, the back patches and via-hole structure also help in widening the impedance bandwidth. A UWB antenna with dual

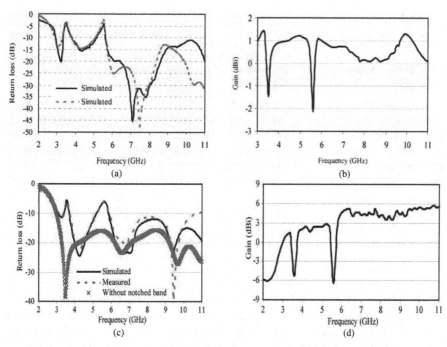

**FIGURE 3.4**
Simulated and measured results for stub and parasitic element loaded dual notch UWB antennas (a) Simulated and measured return loss [50], (b) Gain versus frequency graph [50], (c) Simulated and measured return loss [52], and (d) Gain versus frequency graph [52].

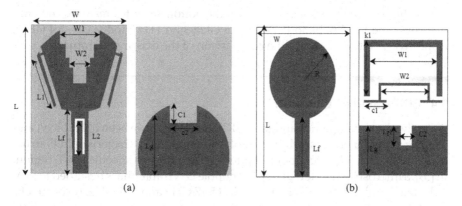

**FIGURE 3.5**
Different types of stub and parasitic element loaded dual notch antennas [re-drawn] (a) V-shaped Antenna [51], and (b) Inverted U-shaped antenna [53].

**TABLE 3.2**

Parasitic-Elements/Stubs-Loaded Dual Band-notched UWB Antennas: A Comparison

| Ref. | Design Approach | Notch Frequency/ Notch-Band (GHz) | Notch-Band Applications | Overall Geometry (mm³) |
|---|---|---|---|---|
| [48] | SIR | 3.2–3.7 & 5–6 | WiMAX, WLAN | 35.7×25×1 (= 892) |
| [49] | Parasitic slits | 3.5 & 5.51 | WiMAX, WLAN | 22×24×1.6 (= 844) |
| [50] | Parasitic strips | 3.3–3.8 & 5.1–5.6 | WiMAX, WLAN | 14.5×14.75×1.6 (= 342) |
| [51] | Parasitic stubs | 3.17–3.85 & 7.9–9.1 | WiMAX, X-band | 23×32×1.62 (= 1192) |
| [52] | Tri-arm resonator | 3.5 & 5.5 | WiMAX, WLAN | 20.5×29×1.6 (= 951) |
| [53] | Parasitic resonator | 5–5.4 & 7.8–8.4 | WLAN, X-band | 32×24×0.762 (= 585) |
| [54] | I/E stubs | 3–3.9 & 5–6 | WiMAX, WLAN | 40×35×1 (= 1400) |
| [55] | Parasitic patch | 3.3–3.8 & 5–6 | WiMAX, WLAN | 25×29×1.6 (= 1160) |
| [56] | T-shaped stubs/U-shaped strips | 3.3–4.0 & 5.05–5.9 | WiMAX, WLAN | 36×26×0.8 (= 748) |
| [57] | T-shaped parasitic strip | 3.22–4.06 & 4.84–5.96 | WiMAX, WLAN | 30×12×0.75 (= 270) |
| [58] | Rectangular resonator | 5.72–5.82 & 8.02–8.4 | WLAN, X-band | 40×40×0.762 (= 1219) |

notched bands is proposed and investigated in [56]. The antenna consists of a square patch and a modified grounded plane. To realize dual notched bands of 3.3–4.0 GHz (WiMAX band) and 5.05–5.90 GHz (WLAN band), a T-shaped stub embedded in the square slot of the radiation patch and a pair of U-shaped parasitic strips beside the feed line are used. Abedian et al. [57] have designed a simple monopole UWB antenna with band-notched function at 3.22–4.06 GHz and 4.84–5.96 GHz, which has been accomplished by embedding a stub that is located at the hollow center of a U-shaped feedline and an inverted T-shaped parasitic strip at the back plane of the antenna. A new CPW-fed UWB planar monopole antenna with dual band-reject characteristics is proposed in [58]. Two resonators of different lengths are employed at the bottom layer to create two notches that cover the WLAN band (5.725–5.825 GHz) and ITU band (8.025–8.4 GHz), respectively. Table 3.2 shows the comparison of the existing literature.

## 3.4 EBG-Loaded Geometries

Recently, electromagnetic bandgap (EBG) structures have been introduced and used for different purposes. By virtue of their bandgap characteristic EBG structures are extensively used to reject interference within a specific UWB region. In [59], a microstrip line-fed planar monopole UWB antenna is designed to provide dual band-notched characteristics at 5.17 GHz and 7.97 GHz, respectively. The filtering characteristics are obtained by placing a mushroom-type EBG unit cell near the feed line. Moreover, the proposed antenna provides satisfactory gain with a stable radiation pattern. A novel design technique has been introduced by Wang et al. [60] which combines mushroom-shaped EBG structures and a slot together for producing dual notched-band characteristics at

WLAN (4.8–5.9 GHz) and X-band downlink satellite communication band (7.1–7.8 GHz). It is also observed that on implementation of the slot, the antenna not only creates its own notched band but also enhances the filtering performance at the other notched band generated by the EBG structures. A good wideband radiation performance from 2.64 GHz to 12.9 GHz has been achieved. A new slitted EBG structure has been designed and implemented to filter out interfering frequencies within the UWB frequency range. The EBG structure placed near the feed line of circular patch UWB antenna leads to the generation of notches for applications in WiMAX and WLAN bands [61].

The UWB disc monopole antenna with crescent-shaped slot for double band-notched features is presented in [62]. Defected ground compact electromagnetic bandgap (DG-CEBG) designs are used to accomplish band-notches in 3.3–3.6 GHz for WiMAX band and 5–6 GHz for WLAN bands. The EBG structures are etched over the ground plane to achieve compactness of around 46% and 50% in comparison with mushroom-type EBG structure. The simulated and measured VSWR curve is shown in Figure 3.6(a).

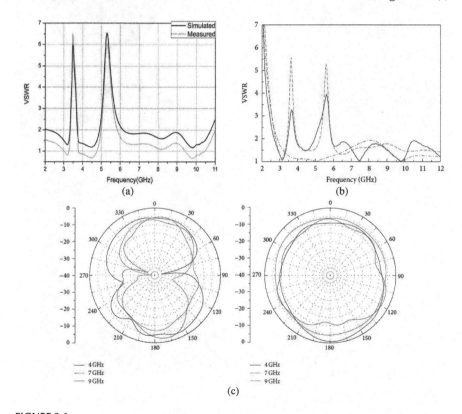

**FIGURE 3.6**
Simulated and measured results for EBG loaded UWB antennas (a) Simulated and measured VSWR curve [62], (b) Simulated return loss [68], and (c) Simulated radiation pattern for different frequencies [68].

Cheng et al. [63] have presented a 2.4–5 GHz WLAN dual band monopole with wideband EBG ground plane. The EBG structures are placed in such a way that they reduce the ground antenna influence and improve the radiation performance. Peng et al. [64] have designed a UWB antenna with wide rectangular notched-band functions at WLAN band (5.150–5.825 GHz) and X-band downlink satellite communication band (7.1–7.76 GHz). The rectangular notched-band design is realized by placing dual mushroom-type EBG structures on the CPW feeding line. In [65], a compact

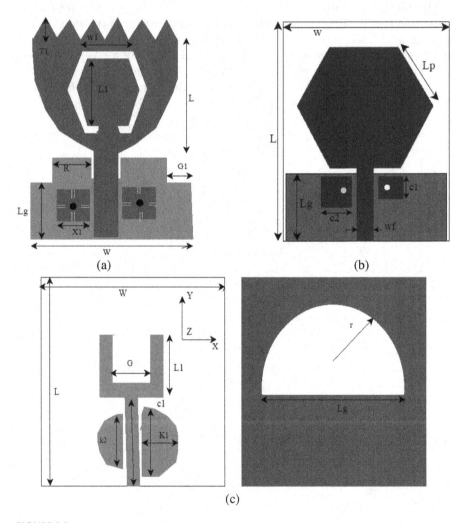

**FIGURE 3.7**
Different types of EBG loaded UWB dual notched UWB antennas [re-drawn] (a) Modified Mushroom type EBG structure [65], (b) Mushroom type EBG structure [67], and (c) Modified Mushroom type EBG structure [69].

dual band-notched UWB monopole antenna is presented for WiMAX and X-band satellite applications. Dual band-rejection characteristics are obtained by engraving the complementary hexagonal spit ring meta-material on the radiating element and two modified mushrooms-like EBG structure above the ground plane in the vicinity of the feed line as depicted in Figure 3.7(a). Based on the mushroom-type EBG structures, a UWB antipodal Vivaldi antenna with high-Q stopband characteristics is designed [66]. The band-notched characteristics at 3.6–3.9 GHz and 5.6–5.8 GHz are obtained by using two pairs of EBG cells along the feed line. Furthermore, the proposed antenna offers flexibility in terms of controlling notched frequencies. A printed hexagonal monopole antenna with dual band rejection is shown in Figure 3.7(b). The dual band rejection at 3.3–3.8 GHz and 5.13–5.88 GHz are realized by placing a pair of mushroom-type EBG structures near the feedline. Moreover, the proposed UWB antenna also exhibits constant group delay and linear phase in the pass band [67]. In [68], a modified EBG structure has been implemented to generate dual notched bands (3.5–5.5 GHz) whose performance is shown in Figure 3.6(b) and (c). A novel UWB wide-slot antenna with dual band-notched characteristics is proposed and the corresponding geometry is shown in Figure 3.7(c). The antenna consists of a fork-shaped feed structure and a ground plane with a quasi-semi-circular slot. The notched bands at 5.2–5.8 GHz for WLAN band are realized by the introduction of a modified mushroom-type semi-circular electromagnetic bandgap structure [69]. Table 3.3 shows the comparison of the existing literature.

**TABLE 3.3**

EBG-Loaded Dual Band-notched UWB Antennas: A Comparison

| Ref. | EBG Structure Used | Notch Frequency/ Notch Band (GHz) | Dual Notch Band | Overall Geometry (mm³) |
|------|--------------------|-----------------------------------|-----------------|------------------------|
| [59] | Mushroom EBG | 5.17 and 7.97 | WLAN, X-band | 38×40×1.6 (= 2432) |
| [60] | Mushroom EBG | 4.8–5.9 and 7.1–7.8 | WLAN, X-band | 31×30.5×1.57 (= 1484) |
| [61] | Slitted EBG | 3.5 and 5.5 | WiMAX, WLAN | 30×40×1.6 (= 1920) |
| [62] | DG-CEBG | 3.3–3.6 and 5–6 | WiMAX, WLAN | 42×50×1.6 (= 3360) |
| [63] | Mushroom EBG | 2.4 and 5.0 | Bluetooth, WLAN | 75×75×0.8 (= 4500) |
| [64] | Mushroom EBG | 5.15–5.825 and 7.10–7.76 | WLAN, X-band | 50×48×1 (= 2400) |
| [65] | Modified Mushroom EBG | 3.5 and 7.6 | WiMAX, X-band | 30×11.4×1.6 (= 547) |
| [66] | Mushroom EBG | 3.6–3.9 and 5.6–5.8 | WiMAX, WLAN | 66.3×66.3×0.813 (= 3573) |
| [67] | Mushroom EBG | 3.3–3.8 and 5.13–5.88 | WiMAX, WLAN | 32×56×1.59 (= 2849) |
| [68] | M-EBG | 5.2 and 5.8 | WLAN | 38×40×1.0 (= 1520) |
| [69] | Modified mushroom EBG | 4.55–5.46 and 5.68–6.35 | WLAN | 35×35×1 (= 1225) |

## 3.5 Summary

Due to the increasing demand for high data and low power consumption, researchers are looking for solutions that produce optimal results. UWB systems provide a better solution for short-range communications with high data rate and low power consumption. In this chapter, designing of UWB antenna with dual band-notched functions has been discussed. From the above state of the art, it is found that the UWB antennas need to be compact for implementation in the modern wireless communication systems. However, it is also necessary to have a UWB antenna with high rejection characteristics. Various techniques have been discussed here for achieving dual band-rejection characteristics. Furthermore, band-notched behavior based on EBG structures is also discussed, but the literature is very limited.

## References

[1] FCC, Washington, DC, Federal Communications commission revision of part 15 of the commission's rules regarding ultra-wideband transmission systems. First reported Order FCC: 02.V48, 2002.

[2] A. M. Abbosh, "Design of a CPW-fed band-notched UWB antenna using a feeder-embedded slot-line resonator," International Journal of Antennas and Propagation, vol. 2008, pp. 1–5, 2008.

[3] A. Ghosh, M. Gangopadhyay, "Bandwidth optimization of microstrip patch antenna –A basic overview," International Journal on Recent and Innovation Trends in Computing and Communication, vol. 6, no. 2, pp. 2321–8169, 2018.

[4] J. Liang, Antenna Study and Design for Ultra Wideband Communication Applications, United Kingdom: University of London, 2006.

[5] H. G. Schantz, "Introduction to Ultra-Wide Band Antennas," In: Proceedings from IEEE Conference on Ultra Wideband System Technologies, Reston, USA, pp. 1–9, 2003.

[6] F. Mouhouche, A. Azrar, M. Dehmas, K. Djafri, "Compact Dual-Band Reject UWB Monopole Antenna using EBG Structures," In: International Conference on Electrical Engineering-Boumerdes (ICEE-B), Boumerdes, pp. 1–5, 2017.

[7] R. Shi, X. Xu, J. Dong, Q. Luo, "Design and analysis of a novel dual band-notched UWB antenna," International Journal of Antennas and Propagation, vol. 2014, Article ID 531959, p. 10, 2014.

[8] D. Rosepriya, Kayalvizhi, "Compact Design of Slot Antenna with Band-notched Function for UWB Application," In: International Conference on Electronics and Communication Systems (ICECS), Coimbatore, pp. 1012–1016, 2015.

[9] J. Liu, S. Gong, Y. Xu, X. Zhang, C. Feng, N. Qi, "Compact printed ultra-wideband monopole antenna with dual band-notched characteristics," Electronics Letters, vol. 44, no. 12, pp. 710–711, 2008.

[10] Q. Chu, Y. Yang, "A compact ultrawideband antenna with 3.4/5.5 GHz dual band-notched characteristics," IEEE Transactions on Antennas and Propagation, vol. 56, no. 12, pp. 3637–3644, 2008.

[11] Y. Zhang, F. S. Zhang, R. Zou, Y. B. Yang, F. Ding, "Design of an Novel Ultra-wideband Antenna with Dual Band-notched Characteristics," In: IEEE CIE International Conference on Radar, Chengdu, pp. 1179–1181, 2011.

[12] Z. Hong, Y. Jiao, B. Yang, W. Zhang, "A Dual Band-notched Antenna for Ultra-wideband Applications," In: IEEE International Conference on Microwave Technology & Computational Electromagnetics, Beijing, pp. 200–202, 2011.

[13] Y. S. Li, X. D. Yang, C. Y. Liu, T. Jiang, "Compact CPW-fed ultra-wideband antenna with dual band-notched characteristics," Electronics Letters, vol. 46, no. 14, pp. 967–968, 2010.

[14] D. Yu, G. Xie, Z. Liao, W. Zhai, "Novel dual band-notched omni-directional UWB antenna on double substrates crossing," IEEE International Symposium on Microwave, Antenna, Propagation and EMC Technologies for Wireless Communications, Beijing, pp. 95–97, 2011.

[15] R. Azim, M. S. Alam, N. Misran, A. T. Mobashsher, M. T. Islam, "Compact Planar Antenna with Dual Band-notched Characteristics for UWB Applications," In: IEEE International Conference on Space Science and Communication (IconSpace), Penang, pp. 269–272, 2011.

[16] W. T. Li, Y. Q. Hei, W. Feng, X. W. Shi, "Planar antenna for 3G/Bluetooth/WiMAX and UWB applications with dual band-notched characteristics," IEEE Antennas and Wireless Propagation Letters, vol. 11, pp. 61–64, 2012.

[17] E. Tammam, K. Yoshitomi, A. Allam, M. El-Sayed, H. Kanaya, K. Yoshida, "A Highly Miniaturized Planar Antenna with Dual Band-notched Using Two Slot Types for UWB Wireless Communications," In: Asia Pacific Microwave Conference Proceedings, Kaohsiung, pp. 726–728, 2012.

[18] F. Zhao, H. Tang, C. Zhao, X. Gao, P. Zhuo, F. Zhang, "Design of novel dual band-notched disk monopole antennas," International Workshop on Microwave and Millimeter Wave Circuits and System Technology, Chengdu, pp. 1–4, 2012.

[19] M. Mehranpour, J. Nourinia, C. Ghobadi, M. Ojaroudi, "Dual band-notched square monopole antenna for ultra wideband applications," IEEE Antennas and Wireless Propagation Letters, vol. 11, pp. 172–175, 2012.

[20] W. Aiting, G. Boran, "A compact CPW-fed UWB antenna with dual band-notched characteristics," International Journal of Antennas and Propagation, vol. 2013, Article ID 594378, p. 7, 2013.

[21] X. Gao, H. W. Yang, H. W. Lai, K. K. So, H. Wong, Q. Xue, "CPW-fed slot antenna with dual band-notched characteristic for UWB application," International High Speed Intelligent Communication Forum, Nanjing, Jiangsu, pp. 1–3, 2012.

[22] N. Ojaroudi, M. Ojaroudi, N. Ghadimi, "Dual band-notched small monopole antenna with novel W-shaped conductor backed-plane and novel T-shaped slot for UWB applications," IET Microwaves, Antennas & Propagation, vol. 7, no. 1, pp. 8–14, 11 January 2013.

[23] D. Kulkarni, J. Varavdekar, "Ultra-Wideband Microstrip Antenna with Dual band-Notched Characteristics Using SRR," In: International Conference on Advances in Recent Technologies in Communication and Computing (ARTCom2012), Bangalore, India, pp. 86–89, 2012.

[24] N. M. Awad, M. K. Abdelazeez, "Bluetooth/UWB Circular Patch Antenna with Dual Band Notches," In: IEEE Jordan Conference on Applied Electrical Engineering and Computing Technologies (AEECT), Amman, pp. 1–4, 2013.

[25] Y. E. Jalil, C. K. Chakrabarty, B. Kasi, "A compact ultra wideband antenna with band-notched design," IEEE Second International Symposium on Telecommunication Technologies (ISTT), Langkawi, pp. 408–412, 2014.

[26] M. Ojaroudi, N. Ojaroudi, N. Ghadimi, "Dual band-notched small monopole antenna with novel coupled inverted U-ring strip and novel fork-shaped slit for UWB applications," IEEE Antennas and Wireless Propagation Letters, vol. 12, pp. 182–185, 2013.

[27] N. M. Awad, M. K. Abdelazeez, "New UWB Antenna with Inverted F and U Shape Slots to Reject WLAN and X-band Applications," In: IEEE GCC Conference and Exhibition (GCC), Doha, pp. 80–83, 2013.

[28] M. Q. Mohammed, A. M. Fadhil, "New Compact Design of Dual Notched Bands UWB Antenna with Slots in Radiating and Feeding Elements," In: IEEE Student Conference on Research and Development, Putrajaya, 374–379, 2013.

[29] J. Shen, C. Lu, J. Zhang, "Heart-shaped Dual Band-notched UWB Antenna," In: Asia-Pacific Conference on Antennas and Propagation, Harbin, pp. 487–490, 2014.

[30] S. Kalraiya, H. S. Singh, G. K. Pandey, A. K. Singh, M. K. Meshram, "CPW-fed Fork Shaped Slotted Antenna with Dual-band Notch Characteristics," In: Students Conference on Engineering and Systems, Allahabad, pp. 1–5, 2014.

[31] X. Gao, Y. Li, Y. Kong, T. Jiang, "A slot-shaped UWB monopole antenna with frequency rejections at WLAN and WiMAX bands," International Microwave Workshop Series on RF and Wireless Technologies for Biomedical and Healthcare Applications (IMWS-BIO), Taipei, pp. 122–123, 2015.

[32] A. Yadav, D. Sethi, S. Kumar, S. L. Gurjar, "L and U Slot Loaded UWB Microstrip Antenna: C-Band/WLAN Notched," In: IEEE International Conference on Computational Intelligence & Communication Technology, Ghaziabad, pp. 380–384, 2015.

[33] M. Sahoo, S. Pattnaik, S. Sahu, "Design of Compact UWB Hexagonal Monopole Antenna with Frequency Notch Characteristics," In: International Conference on Circuits, Power and Computing Technologies [ICCPCT-2015], Nagercoil, pp. 1–4, 2015.

[34] K. Srivastava, A. Kumar, B. K. Kanaujia, "Integrated 23-Cm and UWB Antenna with Dual Notched Characteristics," In: IEEE MTT-S International Microwave and RF Conference (IMaRC), Hyderabad, pp. 352–355, 2015.

[35] A. Garg, D. Kumar, P. K. Dhaker, I. B. Sharma, "A Novel Design Dual Band-notch Small Square Monopole Antenna with Enhanced Bandwidth for UWB Application," In: International Conference on Computer, Communication and Control (IC4), Indore, pp. 1–5, 2015.

[36] A. S. Abbas, M. K. Abdelazeez, "A dual band notch planar SWB antenna with two vertical sleeves on slotted ground plane," IEEE International Symposium on Antennas and Propagation (APSURSI), Fajardo, pp. 2131–2132, 2016.

[37] R. K. Thakur, R. K. Pandey, T. Shanmuganantham, "CPW-Fed Bull Head Shaped UWB Antenna for WiMAX/WLAN with Band-Notched Characteristics," In: International Conference on Circuit, Power and Computing Technologies (ICCPCT), Nagercoil, pp. 1–4, 2016.

[38] A. Saxena, R. P. S. Gangwar, "A Compact UWB Antenna with Dual Band-notched at WiMAX and WLAN for UWB Applications," In: International Conference on Electrical, Electronics, and Optimization Techniques (ICEEOT), Chennai, pp. 4381–4386, 2016.

[39] X. Guan et al., "Novelultra-wideband antenna with dual-band rejection characteristic for wearable applications," IEEE International Workshop on Electromagnetics: Applications and Student Innovation Competition (iWEM), Nanjing, pp. 1–3, 2016.

[40] A. Das, S. Pahadsingh, S. Sahu, "Compact Microstrip Fed UWB Antenna with Dual Band Notch Characteristics," In: International Conference on Communication and Signal Processing (ICCSP), Melmaruvathur, pp. 0751–0754, 2016.

[41] N. Manshouri, A. Yazgan, M. Maleki, "A Microstrip-fed Ultra-wideband Antenna with Dual Band-notch Characteristics," In: International Conference on Telecommunications and Signal Processing (TSP), Vienna, pp. 231–234, 2016.

[42] G. Singh, U. Singh, "Dual Band Rejected Low Profile Planar Monopole Antenna for UWB Application," In: International Conference on Automation, Computational and Technology Management (ICACTM), London, United Kingdom, pp. 534–538, 2019.

[43] S. Priyadharshini, C. Mohan, S. Esther Florence, "Design of Compact-Circular Monopole UWB Antenna with Dual Notch Bands," In: International Conference on Innovations in Information and Communication Technology (ICIICT), Chennai, India, pp. 1–5, 2019.

[44] S. Gogikar, S. Chilukuri, "A Compact Wearable Textile Antenna with Dual Band-Notched Characteristics for UWB Applications," In: IEEE-APS Topical Conference on Antennas and Propagation in Wireless Communications (APWC), Granada, Spain, pp. 426–430, 2019.

[45] M. S. Soliman, M. O. Al-Dwairi, A. Y. Hendi, Z. Alqadi, "A Compact Ultra-Wideband Patch Antenna with Dual Band-Notch Performance for WiMAX/WLAN Services," In: IEEE Jordan International Joint Conference on Electrical Engineering and Information Technology (JEEIT), Amman, Jordan, pp. 831–834, 2019.

[46] S. Kundu, "An Ultra-wideband Dual Frequency Notched Circular Monopole Antenna for Ground Penetrating Radar application," In: URSI Asia-Pacific Radio Science Conference (AP-RASC), New Delhi, India, pp. 1–4, 2019.

[47] H. B. Bapat, V. N. Kamble, M. Tamrakar, "Dual Band Rejected UWB Antenna using Shorted CSRR," In: International Conference on Vision Towards Emerging Trends in Communication& Networking (ViTECoN), Vellore, India, pp. 1–3, 2019.

[48] J. Dong, G. Fu, J. Zhao, X. Wang, "Dual Band-notched UWB Antenna with Folded SIRs," In: International Conference on Microwave and Millimeter Wave Technology (ICMMT), Shenzhen, pp. 1–3, 2019

[49] R. Azim, M. T. Islam, A. T. Mobashsher, "Design of a dual band-notch UWB slot antenna by means of simple parasitic slits," IEEE Antennas and Wireless Propagation Letters, vol. 12, pp. 1412–1415, 2013.

[50] R. Azim, M. T. Islam, A. T. Mobashsher, N. Misran, B. Yatim, "Compact Printed Ultra-wideband Antenna with Dual Band-notch Characteristics," In: International Conference on Computer and Information Technology (ICCIT), Dhaka, pp. 117–121, 2011.

[51] F. Guichi, M. Challal, "Compact UWB Monopole Antenna with WiMAX/ITU Band Notch Characteristics," In: International Conference on Electrical Engineering-Boumerdes (ICEE-B), Boumerdes, pp. 1–4, 2017.

[52] R. Azim, M. T. Islam, A. T. Mobashsher, "Dual band-notch UWB antenna with single tri-Arm resonator," IEEE Antennas and Wireless Propagation Letters, vol. 13, pp. 670–673, 2014.

[53] D. Yadav, P. M. Abegaonkar, K. S. Koul, V. Tiwari, D. Bhatnagar, "A compact dual band-notched UWB circular monopole antenna with parasitic resonators," AEU-International Journal of Electronics and Communications, vol. 84, pp. 313–320, 2018.

[54] W. Hu, J. Zhang, F. Hu, "Design of an ultra-wideband antenna with dual band-notched characteristic," International Symposium on Antennas, Propagation and EM Theory, Guangzhou, pp. 57–59, 2010.

[55] V. H. Nguyen, T. Maeda, "A Compact UWB Antenna with Dual Band-notched Characteristics using Via-hole Structure," In: International Conference on Advanced Technologies for Communications (ATC), Da Nang, pp. 279–282, 2011.

[56] W. Jiang, W. Che, "A novel UWB antenna with dual notched bands for WiMAX and WLAN applications," IEEE Antennas and Wireless Propagation Letters, vol. 11, pp. 293–296, 2012.

[57] M. Abedian, S. K. A. Rahim, S. Danesh, S. Hakimi, L. Y. Cheong, M. H. Jamaluddin, "Novel design of compact UWB dielectric resonator antenna with dual-band-rejection characteristics for WiMAX/WLAN bands," IEEE Antennas and Wireless Propagation Letters, vol. 14, pp. 245–248, 2015.

[58] H. A. Atallah, A. B. Abdel-Rahman, K. Yoshitomi, R. K. Pokharel, "Design of dual band-notched CPW-fed UWB planar monopole antenna using microstrip resonators," Progress in Electromagnetics Research Letters, vol. 59, pp. 51–56, 2016.

[59] A. Ghosh, G. Sen, M. Kumar, S. Das, "Design of UWB antenna integrated with dual GSM functionalities and dual notches in the UWB region using single branched EBG inspired structure," IET Microwaves, Antennas & Propagation, vol. 13, no. 10, pp. 1564–1571, 2019.

[60] Y. Wang, T. Huang, D. Ma, P. Shen, J. Hu, W. Wu, "Ultra-wideband (UWB) Monopole Antenna with Dual Notched Bands by Combining Electromagnetic-Bandgap (EBG) and Slot Structures," In: IEEE MTT-S International Microwave Biomedical Conference (IMBioC), Nanjing, China, pp. 1–3, 2019.

[61] M. Ghahremani, C. Ghobadi, J. Nourinia, M. S. Ellis, F. Alizadeh, B. Mohammadi, "Miniaturised UWB antenna with dual-band rejection of WLAN/WiMAX using slitted EBG structure," IET Microwaves, Antennas & Propagation, vol. 13, no. 3, pp. 360–366, 2019.

[62] N. Jaglan, S. D. Gupta, B. K. Kanaujia, S. Srivastava, E. Thakur, "Triple band notched DG-CEBG structure based UWB MIMO/diversity antenna," Progress in Electromagnetics Research C, vol. 80, pp. 21–37, 2018.

[63] C. Lele et al., "Dual-band Monopole Antenna with Multi-band EBG Ground Plane for 2.4/5 GHz WLAN Applications," In: IEEE International Conference on Microwave and Millimeter Wave Technology (ICMMT), Beijing, 2016, pp. 734–736.

[64] L. Peng, B. Wen, X. Li, X. Jiang, S. Li, "CPW fed UWB antenna by EBGs with wide rectangular notched-band," IEEE Access, vol. 4, pp. 9545–9552, 2016.

[65] F. Mouhouche, A. Azrar, M. Dehmas, K. Djafri, "Compact Dual-band Reject UWB Monopole Antenna Using EBG Structures," In: International Conference on Electrical Engineering-Boumerdes (ICEE-B), Boumerdes, pp. 1–5, 2017.

[66] K. A. Alshamaileh, M. J. Almalkawi, V. K. Devabhaktuni, "Dual band-notched microstrip-fed vivaldi antenna utilizing compact EBG structures," International Journal of Antennas and Propagation, vol. 2015, Article ID 439832, p. 7,2015.

[67] T. Mandal, S. Das, "Design of dual notch band UWB printed monopole antenna using electromagnetic-bandgap structure," Microwave Optical Technology Letter, vol. 56, pp. 2195–2199, 2014.

[68] H. Liu, Z. Xu, "Design of UWB monopole antenna with dual notched bands using one modified electromagnetic-bandgap structure," The Scientific World Journal, vol. 2013, Article ID 917965, pp. 1–9, 2013.

[69] Y. Yang, Z. Y. Yin, F. A. Sun, H. S. Jing, "Design of a UWB wide slot antenna with 5.2/5.8 GHz dual notched bands using modified electromagnetic bandgap structures," Microwave Optical Technology Letter, vol. 54, pp. 1069–1075, 2012.

# 4

## Multi Band-Notched UWB Antennas

### 4.1 Introduction

In this chapter, ultra-wide band (UWB) antennas with triple/quad notched-band characteristics have been presented and discussed. As a key component of the UWB systems, antennas have drawn substantial attention among the researchers. Generally, UWB antennas exhibit different impressive features including wide bandwidth, low power consumption, high data rates, and easy fabrication [1–3]. However, in practical applications, the UWB antennas face frequency band-overlapping challenges. Some narrowband frequencies such as WiMAX (3.3–3.7 GHz), WLAN (5.15–5.35 GHz and 5.725–5.825 GHz), C-band satellite (3.77–4.2 GHz for downlink, 5.9–6.4 GHz for uplink), and X-band ITU (8.025–8.4 GHz band) already exist within the UWB region and cause EM interference [4–8]. In order to reduce the interference, several band-notching techniques such as etching of slots on the radiating patch or the ground plane [9–41], introducing multiple parasitic elements and stubs [42–51] have been adopted till date. Moreover, in recent days, electromagnetic bandgap (EBG) structures are also being used for generating notch-band characteristics in UWB antennas [52–55].

### 4.2 Slotted Geometries

The most common method for generating notched-band characteristics in UWB antennas is by slotting the radiator or ground plane with different structures. Works proposed in [9–41] have used different types of slots for generating notched-band characteristics. In [9], a novel semi-circle-shaped aperture UWB antenna with triple band-notched characteristics is proposed. Thus, by etching one complementary split-ring resonator inside a circular patch and interrogation-shaped defected ground structure, tri-band rejection properties have been realized for WiMAX/WLAN bands. Li et al. [10] have designed a CPW-fed circular ring UWB monopole antenna with tri-notched band function. Triple notched-band characteristics at 5.2–5.9 GHz, 4.0–4.4 GHz, and 7.8–8 GHz have been obtained by using two folded slots and a T-shaped tuning stub in the inner portion of a circular ring. Furthermore, by adjusting the length of the slots and the dimensions of the T-shaped tuning stub, notch band properties can be controlled. A CPW-fed UWB-printed monopole antenna with triple band-notched characteristics has been presented in [11]. By implementing quarter-wavelength slots above the radiating patch, dual band-notched characteristics in 3.3–3.6 GHz for WiMAX and 5.15–5.825 GHz for WLAN have been accomplished. Furthermore, by inserting an open-loop band-notched resonator on the back of the substrate, third band-notched in 7.25–7.75 GHz for X-band satellite communication systems has been realized. The proposed structure is compact in size and has good stopband characteristics.

A novel UWB printed planar antenna with triple band-notched characteristics is also proposed in [12]. Dual notched bands of 3.3–3.6 GHz (WiMAX) and 5.1–5.8 GHz (WLAN) are achieved by embedding two arc-shaped slots on the radiating patch. Another notched-band characteristic at 8.0–8.4 GHz is also achieved by employing a curve-shaped DGS on the ground plane. Moreover, the designed antenna is compact in size and has low cost. Triple notched-band characteristics with sharp rejection at WiMAX band (3.3–3.69 GHz), C-band (3.7–4.2 GHz), and WLAN band (5.15–5.35 GHz) are achieved by using a modified half-mode substrate-integrated waveguide (M-HMSIW) cavity with folded defected ground structure slot [13]. Figure 4.1(a) reveals the design of a triple band-notched CPW-fed UWB antenna. The antenna uses three open-ended quarter-wavelength slots to create triple band-notched characteristics for applications in WiMAX, WLAN, and X-band satellite communication systems [14]. A broad impedance bandwidth with good omnidirectional radiation patterns is also achieved. Similarly, triple-band-notched circular-fan-shaped UWB antenna is presented in [15]. The band-notched function within 3.3–3.6 GHz, 5.1–5.8 GHz, and 8–8.4 GHz is realized by employing a curve-shaped DGS on the ground plane with a curve-shaped slot, and a playground-shaped slot on the radiating patch. In [16], planar CPW-fed UWB monopole antenna with triple band-notch characteristics is presented. Notched-band functions at 3.5 GHz and 10.5 GHz have been realized by implementing U-slot above the feed line. An additional notch at 5.5 GHz is also obtained by L-shaped slots etched from the ground plane. A maximum gain of 4.93 dBi is achieved from the proposed structure. To avoid triple band interference, a compact UWB antenna having a square slot over the feed line is presented [17]. The proposed antenna covers bandwidth of 3.1–10.6 GHz with band rejection at WiMAX (3.3–3.88 GHz), WLAN (4.96–6.23 GHz), and ITU (7.9–8.7 GHz), respectively. Rojhani et al. [18] have used an inverted U-shaped slot to obtain notched-band characteristics from 3 GHz to 3.8 GHz and 5.1 GHz to 6.1 GHz for applications in WiMAX and WLAN bands. Furthermore, an additional notched band at 7.8–8.9 GHz has also been realized by implementing T-shaped strips on the back of the substrate

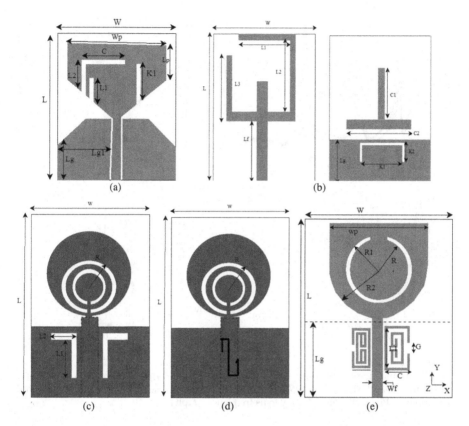

**FIGURE 4.1**
Different types of slot loaded triple-band notched UWB antennas [re-drawn] (a) Open-ended slots [14], (b) U-shaped slot [18], (c) CSRR/L-shaped slots [26], (d) CSRR/S-shape slots [27], and (e) Open-ended slots [37].

(see Figure 4.1(b)). The performance characteristics for the antenna design presented in [18] are shown in Figure 4.2(a). Reddy et al. [19] have designed a novel triple band-notched circular UWB antenna with band-rejection characteristics at WiMAX, WLAN, and C-band. The band-rejection characteristics are obtained by using two concentric G-shaped circular slots etched on the circular patch structure, and a U-shaped slot on the CPW transmission line. In [20], a compact planar UWB antenna with triple band-notched characteristics has been presented. The antenna consists of a lamp post-shaped radiator with a pair of U-shaped slots, and a C-shaped slot for rejection of interference at 3.3–3.7 GHz (WiMAX), 5.15–5.85 GHz (WLAN), and 7.25–8.40 GHz (X-band), respectively. Nearly omni directional radiation pattern and a stable gain over the entire impedance bandwidth except for the three notched bands have also been obtained. Triple band-notched antenna for UWB application is proposed in [21].

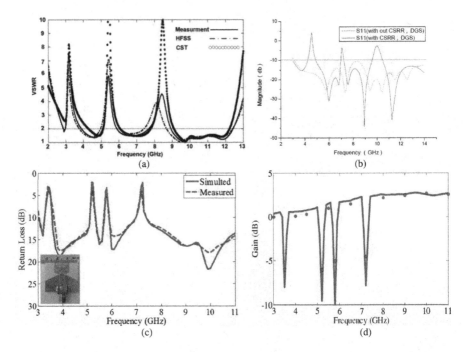

**FIGURE 4.2**
Simulated and measured results for slot loaded triple-band notched UWB antennas (a) Simulated and Measured VSWR plot [18], (b) Simulated $S_{11}$ plot [26], (c) Simulated and Measured $S_{11}$ Plot [34], and (d) Gain versus frequency graph [34].

The notches are obtained by etching a bent and folded slot and a dipole like resonator slots on the antenna radiating patch. The antenna shows band-notched characteristics at the frequencies of 3.55 GHz (WiMAX), 5.75 GHz (WLAN), and 8.27 GHz (X-Band upper band). In [22], Chen et al. have used two pairs of quarter-wavelength long non-uniform slots embedded into the radiating patch and the ground plane to achieve triple notched-band functions at 3.5/5.5/8.1 GHz. Parametric analysis has also been performed to verify the effect of the widths and lengths of the slots. Similarly, triple notch-band planar UWB antenna with a complementary split-ring resonator slot and two L-shaped stubs has been presented in [23]. The notched bands are obtained at 3.3–3.8 GHz WiMAX band, 5.1–5.825 GHz WLAN band, and 7.25–7.75 GHz X-band downlink satellite communication systems. Shaik et al. [24] have designed a new concept of a multi-functional antenna. The proposed antenna consists of a circular patch with a pair of circular SRR along feed line, which exhibits triple notched-band function at 5.24 GHz, 6.4 GHz, and 7.85 GHz, respectively. In [25], a compact, CPW-fed, leaf-shaped UWB antenna with sharp triple notch-bands is presented. The antenna uses two co-directional rectangular split ring-shaped slots on the radiator, and a pair of circular SRR at the back of feed line for realization of notched band at 3.28–3.82 GHz, 5.12–5.4 GHz, and 5.7–6 GHz, which are applicable for rejecting the WiMAX and WLAN bands. Tri-band-notched UWB antenna based on CSRR and L-slots is depicted in Figure 4.1(c). Etching CSRR and a pair of inverted L-shaped slot on radiating patch and ground plane leads to generation of a triple notched-band of 4.1–4.9 GHz, 6.9–7.1 GHz, and 9.3–10.4 GHz, respectively [26]. In addition to this, the simulated $S_{11}$ plot with and without CSRR and DGS is shown in Figure 4.2(b).

Similarly, in [27], triple notched-band functions from 4.1 to 4.9 GHz, 6.8 to 7.1 GHz, and 11.9 to 12.3 GHz have been obtained by implementing a CSRR and S-shaped slot to the feed line, which is shown in Figure 4.1(d). To achieve triple notched-band characteristics, authors have used an electric ring resonator (ERR) over the ground plane of a circular metallic patch [28]. Furthermore, the proposed structure also produces a wide impedance bandwidth of 2.5–12 GHz with controllable notch width by changing the dimensions of the ERR. A compact printed UWB monopole antenna with triple band-notched characteristics is presented in [29]. By etching a split semicircular resonator (SSR) on the radiating patch first, notched band function from 3.3 GHz to 3.6 GHz has been achieved for WiMAX bands. Meanwhile, by adding two capacitively loaded loops (CLLs) close to the micro-strip feed line, two more additional notches have also been realized for applications in WLAN bands. Recently, a coplanar fed quad band-notched UWB antenna is designed [30]. The antenna consists of circular patch with different sizes and shapes of slot in the radiator and ground plane for filtering out the interfering frequencies of 3.12–4.01 GHz, 5.02–5.49 GHz, 5.57–6.03 GHz, and 7–8 GHz, respectively. UWB antenna with quadruple band-notched characteristics is presented in [31]. By incorporating four meander-line slots above the radiator, the proposed antenna exhibits four notches ranging from 3.6 GHz to 4.32 GHz (WiMAX band), 5.17 GHz to 5.32 GHz (WLAN band), 5.80 GHz to 5.93 GHz (WLAN band), and 7.13 GHz to 7.88 GHz (X-band), respectively. Similarly, by implementing open-ended slot in the radiating patch, a single notched-band characteristics within 3.3–3.7 GHz for the WiMAX band is obtained [32]. Meanwhile, by using two pairs of left-handed medium elements near the feed line, triple band-notched properties in bands of 5.15–5.35 GHz, 5.725–5.825 GHz (both WLAN bands), and 7.25–7.75 GHz (X-band) are achieved. In [33], Yu et al. have used a rake-shaped resonator and a pair of L-shaped slots over the patch close to the feed line for achieving quad notched-band properties, which are operated at 3.3–3.6 GHz for WiMAX, 3.9–4 GHz for C-band, 5.6–5.9 GHz for WLAN, and 7.9–8.2 GHz for ITU 8 GHz band. Likewise, by implementing modified V-shaped slots on the rectangular radiating patch, multiband-notches characteristics within 3.3–3.6 GHz (WiMAX band), 5.15–5.35 GHz, 5.725–5.825 GHz (both WLAN bands), and at 7.2 GHz (C-band) have been obtained [34]. The simulated/measured $S_{11}$ plot and its gain versus frequency plot are shown in Figure 4.2(c) and (d).

Abdulhasan et al. [35] have designed quad band-notched UWB antenna. The notch at 3.0 GHz is realized by adding a meander-line strip with the patch.

Similarly, by using F-shaped slots and an inverted diamond-shaped slot on the patch, and J-shaped slots on the ground plane, additional triple notched functions have been accomplished. A hexagonal band-notched UWB antenna based on CSRR, SRR, and DGS have been designed [36]. The rejected band-notched function can be applicable for 3.3–3.6 GHz WiMAX, 5.15–5.35 GHz lower WLAN, 5.7–5.8 GHz upper WLAN, 7.0–7.4 GHz X-band, 8.1–8.5 GHz X-band, and 11.2–12 GHz X-band, respectively. A CPW-fed compact printed monopole antenna with quad band-notched character-istics is presented in [37]. The notched-band characteristics ranging from 3.3 GHz to 3.6 GHz, 5 GHz to 5.4 GHz, 5.7 GHz to 6 GHz, and 7.6 GHz to 8.6 GHz are realized by etching a slot and a split ring resonator on the patch. Furthermore, a rectangular CSRR and open-circuit slots are also being used in the ground plane to obtain notches (see Figure 4.1(d)). Almalkawi and Devabhaktuni [38] have designed a simple multilayered quad band-notched UWB antenna. The quad band-notches are realized by adding closed-loop ring resonators (CLRR). The notches are obtained to reject the interferences of 3.3–3.7 GHz WiMAX band, 4.5–4.8 GHz C-band, 5.15–5.35 GHz, and 5.725–5.825 GHz both WLAN bands. A compact UWB monopole antenna with quad band-notched functions has been proposed in [39]. The suggested structure uses three circular split-ring and a square split-ring to create four notched-band behavior, including WLAN (5.15–5.35 GHz/5.725–5.825 GHz), C-band (3.7–4.2 GHz), and X-band (7.3–7.8 GHz), respectively. Quad band-notched planar UWB antenna is presented in [40]. The suggested antenna comprises a circular patch with a partial truncated ground plane and modified H-shaped resonator. The modified H-shaped resonator near the feedline exhibits dual notched-band characteristics for applications in WiMAX and WLAN bands. Besides, the antenna can produce quad notched-band functions by changing the dimension of H-shaped resonator. Sung [41] has used two U-shaped slots in the radiating patch and another two U-shaped slots into the ground plane for rejecting four interfering frequencies within 3.5 GHz, 5.5 GHz, 9 GHz, and 12.5 GHz, respectively. Finally, Table 4.1 shows the comparison of the existing literature works.

## 4.3 Parasitic-Stub Based Geometries

To generate triple band-notched filtering behavior, a combination of a meander-shaped stub with two rectangular CSRRs near the feedline (see Figure 4.3(a)) and an inverted U-shaped slot on the center of the patch has been proposed [42]. The notched-band characteristics within WiMAX, WLAN, and ITU bands are shown in Figure 4.4(a). A microstrip-fed monopole UWB antenna with triple notch characteristics based on stepped open stubs, CLL resonators, and par-asitic linear segments has been designed [43]. The notched-band functions

**TABLE 4.1**

Comparison of Multi Notched UWB Antenna Using Different Types of Slots

| Ref. | Approach Used | Notched-Bands (GHz) | Notched-Band Applications | Overall Geometry (mm³) |
|---|---|---|---|---|
| Triple Notched-Band | | | | |
| [9] | SRR | (3.3–4)/(5.15–5.4)/ (5.8–6.1) | WiMAX/WLAN | 24×30×1 (= 720) |
| [12] | Curve-shaped DGS | (3.3–3.6)/(5.1–5.8)/ (8.0–8.4) | WiMAX/ WLAN/X-Band | 32×35×0.508 (= 568) |
| [14] | Open-ended quarter wavelength slot | (3.3–3.7)/(5.1–5.8)/ (7.25–7.75) | WiMAX/ WLAN/X-Band | 19×24×1.2 (= 547) |
| [15] | Curve-shaped DGS | (3.3–3.6)/(5.1–5.8)/ (8.0–8.4) | WiMAX/ WLAN/X-Band | 30×24.5×1.6 (= 1176) |
| [17] | Square slots | (3.3–3.8)/(4.96–6.23)/ (7.9–8.7) | Long-term evolution (LTE)-band/C-band/ X-Band | 27×30.5×1.6 (= 1296) |
| [18] | T-shaped element and U-shaped slot | (3.0–3.8)/(5.1–6.1)/ (7.8–8.9) | WiMAX/ WLAN/X-Band | 20×14.5×1 (= 280) |
| [20] | U/C-slot | (3.3–3.7)/(5.15–5.85)/ (7.25–8.4) | WiMAX/ WLAN/X-Band | 30×30×1.6 (= 1440) |
| [22] | Non-uniform slots | 3.5/5.5/8.1 | WiMAX/ WLAN/X-Band | 25×30.2×0.762 (= 575) |
| [26] | CSRR | (4.1–4.9)/(6.9–7.1)/ (9.3–10.4) | C-Bands | 23×30×1 (= 690) |
| [28] | ERR | (3.3–3.6)/(5.15–5.35)/ (5.7–5.8) | WiMAX/WLAN Bands | 50×50×1.52 (= 3800) |
| Multi Notched-Band | | | | |
| [13] | Folded DGS slot | (3.3–3.6)/(3.7–4.2)/ (5.15–5.35), (5.75–5.825) | WiMAX/C-Band/ WLAN | 40×47×1.016 (= 1910) |
| [29] | CSRR/SRR/ DGS | (3.3–3.6)/(5.15–5.35)/ (5.7–5.8) (7–7.4)/ (8.1–8.5)/(11.2–12) | WiMAX/ WLAN/X-Bands | 30×28×0.508 (= 462) |

are obtained for applications in WiMAX band, WLAN band, and X-band. Quadruple band-notch compact UWB antenna is presented in [44]. The suggested structure consists of a UWB semi-elliptical planar monopole with trapezoidal spiral for 2.45-GHz Bluetooth application, and rectangular resonant spiral structures for rejection of frequency bands, i.e., 3.3–3.6 GHz WiMAX, 5.15–5.35 GHz, 5.725–5.825 GHz both WLAN bands and 8 GHz ITU band. Similarly, a compact UWB monopole antenna with triple band-notch characteristic is presented in Figure 4.3(b). For achieving band-notched functions at 3.3–3.6 GHz WiMAX band and 5.15–5.3 GHz WLAN band, 5.7–5.825 GHz bands, the antenna is loaded with non-concentric open-ended rings [45].

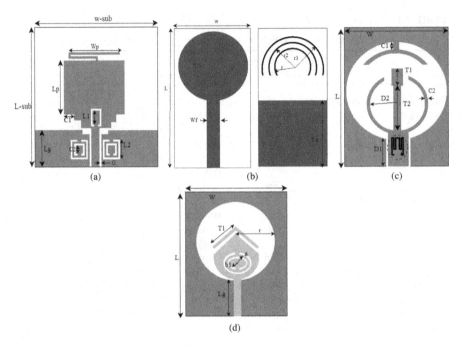

**FIGURE 4.3**
Different types of stubs and parasitic elements loaded triple-band notched UWB antennas [re-drawn] (a) Meander shaped stub [42], (b) Non-concentric open ended rings [45], (c) SIR and a meander line slot [46], and (d) V shaped strip [50].

Figure 4.3(c) reveals an arc-shaped stub, a stepped impedance resonator (SIR), and a meander line slot (MLS) have been used for realizing triple notched-band characteristics at 3.5 GHz (WiMAX), 5.5 GHz (WLAN), and 7.5 GHz (X-band) [46]. Wang and Gao [47] have designed a printed monopole UWB antenna with four rejection bands at 2.6 GHz (TD-LTE), 3.3–4.2 GHz (WiMAX and C-band communication satellite), 5.2–5.8 GHz (WLAN and HIPERLAN/2), and 7.25–7.75 GHz (X-band satellite downlink) by inserting two SRR-shaped slots and two parasitic

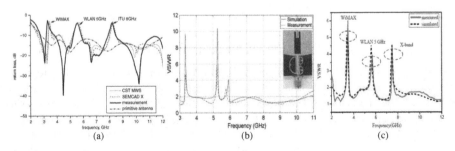

**FIGURE 4.4**
Simulated and measured results for stubs and parasitic elements loaded triple-band notched UWB antennas (a) Simulated and Measured $S_{11}$ plot [42], (b) Simulated and Measured VSWR plot [48], and (c) Simulated and Measured VSWR Plot [50].

meandered ground stubs over the radiator close to the feed line. Two compact, printed, UWB monopole antennas with tri-band-notched characteristics are reported in [48]. By adding three capacitively-loaded loop (CLL) resonators close to the feed line, triple notched-band behavior has been obtained for operation in 3.3–3.6 GHz WiMAX, 5.15–5.35 GHz lower WLAN, and 5.725–5.825 GHz higher WLAN bands, which is shown in Figure 4.4(b). Triple notched-band characteristics have also been realized by implementing a pair of modified capacitance loaded loop resonators near the feed line in [49]. The notches are obtained for applications in 3.4–3.7 GHz for WiMAX, 5.15–5.825 GHz for WLAN, and 7.25–8.395 GHz for X-band satellite communication. A UWB antenna with cone-shaped radiating patch and a wide circularly slot on the ground plane is reported in [50]. Besides this, an elliptical CSRR (ECSRR) and a parasitic V-shaped strip (see Figure 4.3(d)) have been incorporated into the radiating patch, which results in generating triple notched-band behavior at WiMAX, WLAN, and X band satellite communication as depicted in Figure 4.4(c). Moreover, the proposed structure has compact size and stable omnidirectional radiation patterns. Zhu et al. [51] have proposed a multiple band-notch UWB antenna. Quadruple band-notch characteristic has been accomplished by utilizing four pairs of meander lines (MLs) structures near the feed line and over the edge of the ground plane. Table 4.2 shows the comparison of the existing literature.

**TABLE 4.2**

Comparison of Multi Notched UWB Antenna Using Parasitic Stubs/Strips

| Ref. | Design Approach | Notch Band Frequency (GHz) | Notch Band Applications | Overall Geometry (mm³) |
|---|---|---|---|---|
| [42] | Meander stubs | 3.5/5.5/8.2 | WiMAX, WLAN X-band | 40×40×0.812 (= 1299) |
| [43] | Stepped open stubs | 3.6/5.6/8.8 | WiMAX, WLAN X-band | ----- |
| [44] | Non-concentric open stub resonator | (3.3–3.6)/(5.15–5.35)/ (5.75–5.825) | WiMAX, WLAN | 24×17×0.787 (= 321) |
| [45] | Non-concentric ring resonator | (3.3–3.6)/(5.18–5.3)/ (5.7–5.825) | WiMAX/WLAN | 25×18×0.787 (= 354) |
| [46] | Stepped impedance resonator | 3.5/5.5/7.5 | WiMAX/ WLAN/X-band | 32×24×1.6 (= 1228) |
| [47] | Meander ground stubs | 2.6/(3.4–4.2)/ (5.2–5.8)/ (7.25–7.75) | LTE/C-band/ WLAN/X-band | 33×46×0.762 (= 1156) |
| [48] | CLL resonator | (3.3–3.6)/(5.18–5.3)/ (5.7–5.825) | WiMAX/WLAN | 27×34×0.787 (= 722) |
| [49] | Modified CLL resonator | (3.4–3.7)/(5.15–5.825)/ (7.25–8.395) | WiMAX/ WLAN/X-band | ----- |
| [50] | V-shaped strips | 3.5/5.5/7.5 | WiMAX/ WLAN/X-band | 55×55×0.51 (= 1542) |
| [51] | Meander line stubs | 3.14/4.34/5.4/6.4 | WiMAX/ WLAN/C-band | 30×39.3×0.6 (= 702) |

## 4.4 EBG-Loaded Geometries

Recently, the EBG structures are also being used for generating notch characteristics in UWB antennas. Literature works [52–56] have used various EBG structures for generating band-notched functions in UWB antennas. In [52], triple notched-band property has been obtained by using complementary EBG (CEBG) at both sides of the feed line and ring on the ground (see Figure 4.5(a)). Four different configurations of beveled rectangular patch UWB antenna are designed [53]. There by utilizing one, two, and three EBG structures, single-, dual-, and triple-band-rejection properties have been obtained at 3.4 GHz, 5.2 GHz, and 5.8 GHz, respectively. However, to verify the effect of notches, all the structures are simulated individually. A compact microstrip UWB antenna with multiband resonance features is reported in [54]. The band-rejection characteristics at 2.4 GHz, 3.5 GHz, and 5.5 GHz are realized by introducing mushroom-type EBG structure and periodic DGS. Figure 4.5(b) shows the design concept of novel UWB antenna with triple band-notched characteristics. For achieving UWB performance, the authors have used U-shaped dielectric resonator placed at the edge of a rectangular patch. Triple notched-band characteristics at 4.5/5.1/9.1 GHz have been accomplished by implementing two configurations of Archimedean spiral-shaped EBG unit cells, and their simulated performance parameters are shown in Figure 4.6. A comparison is made in Table 4.3, but the literature is very limited.

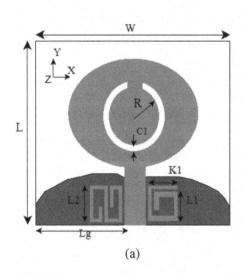

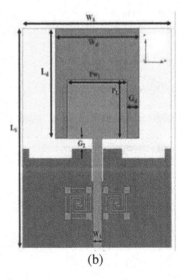

(a)                                          (b)

**FIGURE 4.5**
Different types of EBG loaded multi-notch UWB antennas [re-drawn] (a) Complementary EBG Structure [52], and (b) Archimedean spiral EBG structure [55].

**TABLE 4.3**

EBG-Loaded Band-Notched UWB Antennas: A Comparison

| Ref. | EBG Structures Used | Notch Bands (GHz) | Notch Band Applications | Overall Geometry (mm³) |
|------|---------------------|-------------------|-------------------------|------------------------|
| [52] | CEBG | 2.45/3.55/6 | Bluetooth/WiMAX/C-band | 33×37×1.0 (=1221) |
| [53] | Mushroom EBG | 3.4/5.2/5.8 | WiMAX/WLAN | 22×32×0.8 (=563) |
| [54] | Mushroom EBG | 2.4/3.5/5.5 | Bluetooth/WiMAX/WLAN | 20×20×1.6 (=640) |
| [55] | Archimedean spiral EBG | 4.8/5.1/9.1 | Bluetooth/WLAN | 25×30×0.762 (=571) |

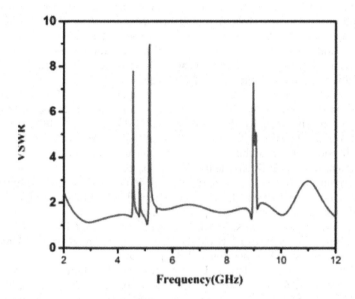

**FIGURE 4.6**
Simulated VSWR performance of Archimedean spiral EBG loaded triple-band notched UWB antenna [55].

## 4.5 Summary

Due to the increasing demand for high data and low power consumption, researchers are looking for solutions that produce optimal results. UWB systems provide a better solution for short-range communications with high data rate and low power consumption. In this chapter, designing of the UWB antenna with triple/multi band-notched functions has been discussed. Various techniques for realization of multi-notch functionality have been discussed here in detail with their respective advantages and disadvantages. Furthermore, the role of EBG structures in achieving multi-notch characteristics have also been investigated here. However, the literature in this case is also very limited and therefore posses immense potential for future research.

## References

[1] FCC, Washington, DC, Federal Communications commission revision of part 15 of the commission's rules regarding ultra-wideband transmission systems. First reported Order FCC: 02.V48, 2002.

[2] P. P. Shome, T. Khan, R. H. Laskar, "A state-of-art review on band-notch characteristics in UWB antennas," International Journal of RF Microwave Computer Aided Engineering, vol. 29, Article ID e21518, 2019.

[3] M. Rahman, S. D. Ko, D. J. Park, "A compact multiple notched ultra-Wide band antenna with an analysis of the CSRR-TO-CSRR coupling for portable UWB applications," Sensors, vol. 17, p. 2174, Sep. 2017.

[4] A. Saxena, S. P. R. Gangwar, "Review on Band-Notching Techniques for Ultra wideband Antenna," In: Proceedings of the International Conference on Nanoelectronics, Circuits & Communication Systems. Lecture Notes in Electrical Engineering, vol. 403. Springer, Singapore, 2017.

[5] T. Saeidi, I. Ismail, W. P. Wen, A. R. H. Alhawari, A. Mohammadi,"Ultrawideband antennas for wireless communication applications," International Journal of Antennas and Propagation, vol. 2019, Article ID 7918765, p. 25, 2019.

[6] R. Cicchetti, E. Miozzi, O. Testa, "Wideband and UWB antennas for wireless applications: A comprehensive review," International Journal of Antennas and Propagation, vol. 2017, Article ID 2390808, p. 45, 2017.

[7] Y. F. Weng, S. W. Cheung, T. I. Yuk, "Triple band-notched UWB antenna using meander ground stubs," In: Proceedings of the 2010 Lughborough Antennas and Propagation Conference, pp. 341–344, 2010.

[8] H. Choudhary, T. Singh, K. A. Ali, A. Vats, P. K. Singh, D. R. Phalswal, V. Gahlaut, "Design and analysis of triple band-notched micro-strip UWB antenna," Journal of Cogent Engineering, vol. 3, pp. 1–13, 2016.

[9] X. Liao, H. Yang, N. Han, Y. Li, "Aperture UWB antenna with triple band-notched characteristics," Electronics Letters, vol. 47, no. 2, pp. 77–79, 2011.

[10] Y. Li, S. Chang, M. Li, X. Yang, "A compact ring UWB antenna with tri-notch band characteristics using slots and tuning stub," IEEE International Symposium on Microwave, Antenna, Propagation and EMC Technologies for Wireless Communications, Beijing, pp. 12–15, 2011.

[11] J. W. Wang, J. Y. Pan, X. N. Ma, Y. Q. Sun, "A Band-Notched UWB Antenna with L-Shaped Slots and Open-Loop Resonator," In: IEEE International Conference on Applied Superconductivity and Electromagnetic Devices, Beijing, pp. 312–315, 2013.

[12] P. Zhuo, H. Tang, X. Gao, F. Zhao, F. Zhang, "Investigation on planar UWB antenna with triple band-notched characteristics," IEEE International Workshop on Electromagnetics: Applications and Student Innovation Competition, Chengdu, Sichuan, pp. 1–3, 2012.

[13] E. Ali-Akbari, M. Azarmanesh, S. Soltani, "Design of miniaturized band-notch ultra-wideband monopole-slot antenna by modified half-mode substrate-integrated waveguide," IET Microwaves, Antennas & Propagation, vol. 7, no. 1, pp. 26–34, 2013.

[14] D. T. Nguyen, D. H. Lee, H. C. Park, "Very compact printed triple band-notched UWB antenna with Quarter-wavelength slots," IEEE Antennas and Wireless Propagation Letters, vol. 11, pp. 411–414, 2012.

[15] J. Tang, M. Chen, Y. Li, "A novel planar UWB antenna with triple band-notched characteristics," Xian, pp. 340–34, 2012.

[16] S. Maiti, N. Pani, A. Mukherjee, "Modal Analysis and Design a Planar Elliptical Shaped UWB Antenna with Triple Band Notch Characteristics," In: International Conference on Signal Propagation and Computer Technology (ICSPCT), Ajmer, pp. 13–15, 2014.

[17] Y. Cai, H. Yang, L. Cai, "Wideband monopole antenna with three band-notched characteristics," IEEE Antennas and Wireless Propagation Letters, vol. 13, pp. 607–610, 2014.

[18] N. Rojhani, M. Akbari, A. Sebak, "Controllable triple band-notched monopole antenna for ultra-wideband applications," IET Microwaves, Antennas & Propagation, vol. 9, no. 4, pp. 336–342, 2015.

[19] B. C. Reddy, E. S. Shajahan, M. S. Bhat, "Design of a Triple Band-notched Circular Monopole Antenna for UWB Applications," In: Eleventh International Conference on Wireless and Optical Communications Networks (WOCN), Vijayawada, pp. 1–5, 2014.

[20] S. Tomar, A. Kumar, "Design of a Novel Compact Planar Monopole UWB Antenna with Triple Band-notched Characteristics," In: International Conference on Signal Processing and Integrated Networks (SPIN), Noida, pp. 56–59, 2015.

[21] X. Hu, W. Yang, S. Yu, R. Sun, W. Liao, "Triple Band-notched UWB Antenna with Tapered Microstrip Feed Line and Slot Coupling for Bandwidth Enhancement," In: International Conference on Electronic Packaging Technology (ICEPT), Changsha, pp. 879–883, 2015.

[22] X. Chen, F. Xu, X. Tan, "Design of a compact UWB antenna with triple notched bands using non uniform width slots," Journal of Sensors, vol. 2017, Article ID 7673168, p. 9, 2017.

[23] S. Doddipalli, A. Kothari, "Compact UWB antenna with integrated triple notch bands for WBAN applications," IEEE Access, vol. 7, pp. 183–190, 2019.

[24] L. A. Shaik, C. Saha, J. Y. Siddiqui,Y. M. M. Antar, "Ultra-wideband monopole antenna for multiband and wideband frequency notch and narrowband applications," IET Microwaves, Antennas & Propagation, vol. 10, no. 11, pp. 1204–1211, 2016.

[25] A. Sohail, S. K. Alimgeer, A. Iftikhar, B. Ijaz, W. K. Kim, W. Mohyuddin, "Dual notch band UWB antenna with improved notch characteristics," Microwave Optical Technology Letter, vol. 60, pp. 925–930, 2018.

[26] W. Liu, T. Jiang, "Design and Analysis of a Tri-band Notched UWB Monopole Antenna," In: Asia-Pacific Conference on Antennas and Propagation (APCAP), Kaohsiung, pp. 385–386, 2016.

[27] W. Liu, T. Jiang, "Design and Analysis of a Tri-band Notch UWB Monopole Antenna," In: Progress in Electromagnetic Research Symposium (PIERS), Shanghai, pp. 2039–2041, 2016.

[28] I. B. Vendik, A. Rusakov, K. Kanjanasit, J. Hong, D. Filonov, "Ultrawideband (UWB) planar antenna with single-, dual- and triple-band notched characteristic based on electric ring resonator," IEEE Antennas and Wireless Propagation Letters, vol. 16, pp. 1597–1600, 2017.

[29] X. Hu, X. Yang, "Tri-band-notched Ultrawideband (UWB) Antenna Using Split Semicircular Resonator (SSR) and Capacitively Loaded Loops (CLL)," In: IEEE International Conference on Communication Problem-Solving (ICCP), Guilin, pp. 186–189, 2015.

[30] J. Xu, C. Du, G. Jin, K. Li, W. Zheng, Z. Zhao, "A coplanar feed quad-band notched UWB antenna," International Workshop on Electromagnetics: Applications and Student Innovation Competition (iWEM), Qingdao, China, pp. 1–2, 2019.

[31] Z. Zhong, G. Huang, T. Yuan, M. Fan, "Compact CPW-Fed UWB Antenna with Quadruple Band-Notched Characteristics," In: Cross Strait Quad-Regional Radio Science and Wireless Technology Conference (CSQRWC), Xuzhou, pp. 1–3, 2018.

[32] J. Zhang, Y. Z. Cai, F. Y. Wang, H. C. Yang, "UWB Band-Notched Monopole Antenna Design Using Left-Handed Materials," In: IEEE International Conference on Applied Superconductivity and Electromagnetic Devices, Beijing, pp. 482–485, 2013.

[33] K. Yu, Y. Li, X. Luo, X. Liu, "A planar UWB antenna with quad notched bands using rake-shaped resonator and L-shaped slots," IEEE International Symposium on Antennas and Propagation (APSURSI), Fajardo, pp. 1815–1816, 2016.

[34] M. Darvish, H. R. Hassani, B. Rahmati, "Compact CPW-fed Ultra Wideband Printed Monopole Antenna with Multi Notch Bands," In: Iranian Conference on Electrical Engineering (ICEE2012), Tehran, pp. 1114–1119, 2012.

[35] R. A. Abdulhasan, R. Alias, K. N. Ramli, "A compact CPW fed UWB antenna with quad band notch characteristics for ISM band applications," Progress In Electromagnetics Research M, vol. 62, pp. 79–88, 2017.

[36] M. J. Pushpa, A. J. Rani, V. Saritha, "A compact hexa-notched antenna using combination of CSRR, SRR and DGS," International Journal of Innovative Technology and Exploring Engineering (IJITEE), vol. 8, no. 8, pp. 2683–2688, 2019.

[37] Y. Cao, J. Wu, H. Yang, "Design of CPW-fed Monopole Antenna with Quadruple Band-notched Function for UWB Application," In: International Conference on Computational Problem-Solving (ICCP), Chengdu, pp. 353–356, 2011.

[38] M. J. Almalkawi, V. K. Devabhaktuni, "Quad band-notched UWB antenna compatible with WiMAX/INSAT/lower-upper WLAN applications," Electronics Letters, vol. 47, no. 19, pp. 1062–1063, 2011.

[39] L. J. Jiang, L.Y. Sheng, L.S. Qun, "Four-Band-Notched UWB Antenna Using Three Circular Split-rings and a Square Split-Ring," In: IEEE International Conference on Communication Problem-Solving, Beijing, pp. 354–356, 2014.

[40] Y. Sung, "Quad band-notched ultra wideband antenna with a modified H-shaped resonator," IET Microwaves, Antennas & Propagation, vol. 7, no. 12, pp. 999–1004, 2013.

[41] S. H. Zainud-Deen, R. A. Al-Essa, S. M. M. Ibrahem, "Ultra wideband printed elliptical monopole antenna with four band-notch characteristics," IEEE Antennas and Propagation Society International Symposium, Toronto, pp. 1–4.

[42] D. Kim, C. Kim, "CPW-fed ultra-wideband antenna with triple-band notch function," Electronics Letters, vol. 46, no. 18, pp. 1246–1248, 2010.

[43] S. Nikolaou, M. Davidovic, M. Nikolic, P. Vryonides, "Triple notch UWB antenna controlled by three types of resonators," IEEE International Symposium on Antennas and Propagation (APSURSI), Spokane, WA, pp. 1478–1481, 2011.

[44] G. S. Reddy, A. Kamma, S. K. Mishra, J. Mukherjee, "Compact Bluetooth/UWB dual-band planar antenna with quadruple band-notch characteristics," IEEE Antennas and Wireless Propagation Letters, vol. 13, pp. 872–875, 2014.

[45] G. S. Reddy, A. Kamma, J. Mukherjee, "Compact Printed Monopole UWB Antenna Loaded with Non-Concentric Open-Ended Rings for Triple Band-Notch Characteristic," In: Asia-Pacific Microwave Conference Proceedings (APMC), Seoul, pp. 221–223, 2013.

[46] Y. Li, W. Zhang, R. Mittra, "A circular wide slot UWB antenna with triple band-notch characteristics," IEEE International Symposium on Antennas and Propagation & USNC/URSI National Radio Science Meeting, Vancouver, BC, pp. 2325–2326, 2015.
[47] N. Wang, P. Gao, "A novel printed UWB and Bluetooth antenna with quad band-notched characteristics," International Workshop on Microwave and Millimeter Wave Circuits and System Technology, Chengdu, pp. 150–153, 2013.
[48] C. Lin, P. Jin, R. W. Ziolkowski, "Single, dual and tri-band-notched ultra wideband (UWB) antennas using capacitively loaded loop (CLL) resonators," IEEE Transactions on Antennas and Propagation, vol. 60, no. 1, pp. 102–109, 2012.
[49] J. Wang, Y. Yin, X. Liu, "Triple band-notched ultra-wideband antenna using a pair of novel symmetrical resonators," IET Microwaves, Antennas & Propagation, vol. 8, no. 14, pp. 1154–1160, 18 11 2014.
[50] J. Zhu, B. Peng, B. Feng, L. Deng, S. Li, "Triple Band-notched Slot Planar Inverted Cone Antenna for UWB Applications," In: Asia-Pacific Microwave Conference (APMC), Nanjing, pp. 1–3, 2015.
[51] Y. F. Weng, S. W. Cheung, T. I. Yuk, "Design of multiple band-notch using meander lines for compact ultra-wide band antennas," IET Microwaves, Antennas & Propagation, vol. 6, no. 8, pp. 908–914, 2012.
[52] F. Zhou, Z. Qian, J. Han, C. Peng, "Ultra-Wideband Planar Monopole Antenna with Triple Band-Notched Characteristics," In: International Conference on Microwave and Millimeter Wave Technology, Chengdu, pp. 438–440, 2010.
[53] L. Peng, C. L. Ruan, "Design and time-domain analysis of compact multi-band-notched UWB antennas with EBG structures," Progress in Electromagnetics Research B, vol. 47, pp. 339–357, 2013.
[54] A. Yadav, S. Goyal, T. Agrawal, R. P. Yadav, "Multiband Antenna for Bluetooth/ZigBee/Wi-Fi/WiMAX/WLAN/X-band Applications: Partial Ground with Periodic Structures and EBG," In: International Conference on Recent Advances and Innovations in Engineering (ICRAIE), Jaipur, pp. 1–5, 2016.
[55] S. Kumar, T. Khan, "EBG-Loaded Dielectric Resonator Antenna for Triple Band-Notch Characteristics," In: URSI Asia-Pacific Radio Science Conference (AP-RASC), New Delhi, India, pp. 1–5, 2019.

# 5

## Band-Notched UWB MIMO Antennas

### 5.1 Introduction

In modern wireless communication systems, high-speed and reliable data transmission without any increment in the bandwidth or the transmitted power is very obligatory. To achieve this goal, multi-input multi-output (MIMO) communication systems are recently being used, and have received more attraction for having capabilities such as robustness against multipath fading, improvement in channel capacity, spatial diversity gain, etc. [1, 2]. This concept is based on using multiple transmitting and receiving antennas to achieve spatial diversity or spatial multiplexing. However, placing multiple elements causes strong coupling between them. Therefore, it is necessary to design a UWB MIMO antenna with high isolation and wider bandwidth [3, 4]. Many studies have been performed on the design and analysis of UWB MIMO antennas. However, in practical applications, designing of UWB antenna causes band overlapping problems with other designated narrowband systems [5–7]. Consequently, to overcome the interfering issues, UWB systems need to have multi band-notched features. The most common way to achieve them is by introducing slots over radiating patch or ground plane. Placing parasitic strips/stubs in close proximity to the radiating element or slits on the radiator patch leads to generation of notches. In recent days, EBG structures are also being used for realizing notched band characteristics in UWB antennas.

### 5.2 Slotted-Based Geometry

In this section, UWB MIMO antennas with single/dual/multiple notched-band characteristics are discussed. Authors in [8–38] have designed various UWB MIMO antennas with slotted geometries for achieving band-rejection properties. In [8], UWB MIMO antenna with band-notched characteristics has been proposed. The notched-band functions at 3.7–5 GHz and 5.85–10.6 GHz are achieved by embedding U-shaped slots on the radiator. Besides this, a truncated ground plane is modified on the reverse side of the board for achieving isolation of less than –15 dB. A compact octagonal-shaped fractal

UWB MIMO antenna is presented in [9]. To realize band-notched character-istics at WLAN band, the suggested antenna in [10] incorporates a C-shaped slot, and an L-shaped stub has been used in the ground plane for obtaining better isolations. Moreover, the proposed structure provides miniaturization using Minkowski-shaped fractal geometry. Similarly, by using C-shaped slot and meander line structure, the suggested design produces notched-band function for WLAN applications with –17.5 dB isolation. Lin et al. [11] have designed 4×4 UWB MIMO antenna. The antenna comprises of a rectangular patch with I and C-Shaped slots for realizing dual notched behaviors in the frequency ranges of WiMAX (3.3–3.8 GHz) and WLAN (5.15–5.35 GHz). To meet high-coupling reduction better than –20 dB, four rectangular and four staircase-shaped stubs are added to the ground plane. Dual band-notched UWB MIMO antenna is designed in [12]. The antenna consists of one-third $\lambda$-open-ended slot, half $\lambda$- parasitic strips, and one-fourth $\lambda$-open-ended slot for realizing notched characteristics in the frequency ranges of WLAN (5.15–5.825 GHz) and WIMAX (3.3–3.7 GHz). Recently, rectangular SRR-slots, as depicted in Figure 5.1(a), have been used for realizing notched-band function

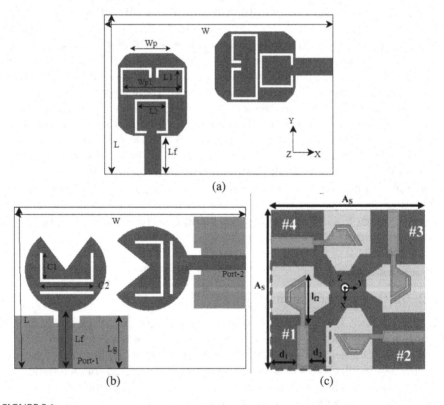

**FIGURE 5.1**
Slot-loaded band-notched UWB MIMO antenna [re-drawn] (a) SRR slot [13], (b) U-shaped slot [14], and (c) Quasi-self-complementary antenna [20].

at WiMAX and WLAN bands [13]. The suggested antenna provides isolation of –15 dB by proposing meander-shaped ground stubs. The analysis of compact UWB MIMO antenna with diversity characteristics and dual band-notched functionality have been carried out in [14]. The antenna consists of a U-shaped slot and a horizontal line slot over the pacman-shaped radiator to reject interference within WiMAX (3.5 GHz) and WLAN (5.5 GHz) (See Figure 5.1(b)). Two-port MIMO antenna with dual band-notched characteristics for UWB applications is presented in [15]. The antenna exhibits dual band-rejection properties by introducing U-shaped slot in the main radiator and horizontal stubs in the ground plane. The notches are obtained at 5.15–5.825 GHz (WLAN) and 7.25–7.75 GHz (X-band satellite downlink). Besides this, the antenna provides an isolation of more than –20 dB with the help of decoupling structure.

By etching two semi-circular slots on the radiating patch of a UWB MIMO antenna, dual band rejections within the WiMAX (3.24–3.87 GHz) and WLAN (4.72–5.64 GHz) bands are achieved in [16]. Furthermore, an isolation of –19.72 dB is achieved by introducing a parasitic stub along with the ground plane. Yu et al. in [17] have proposed two-element UWB MIMO antenna with quad narrow-band frequency rejection characteristics. The notched-band functions within 3.1–3.7 GHz, 4.2–4.7 GHz, 5.6–5.9 GHz, and 7.9–8.3 GHz bands are accomplished by using a pair of L-shaped slots (LSSs) and a pair of square ring short stub loaded resonators (SRSSLRs). Likewise, by etching two L-shaped slots on the ground plane and two anchor-shaped stubs on the radiating patch, dual band-filtering functions for WiMAX and WLAN bands have been obtained [18]. Furthermore, the suggested geometry provides high isolation, which is better than –19.74 dB.

A compact dual-polarized MIMO antenna with dual band rejection characteristics for UWB applications is proposed in [19]. The notched bands at 3.3–4 GHz WiMAX and 5.1–6 GHz WLAN frequencies are obtained by using a quarter-wavelength bent slot and a quarter-wavelength slot over the radiating patch and the ground plane. Meanwhile, the two quasi-self-complementary (QSC) radiating elements provide high mutual coupling reduction. Similarly, based on the QSC elements, four ports UWB MIMO antenna shown in Figure 5.1(c) is designed with single band-rejection at WLAN band [20]. The band-notched feature is realized by using a slit cut on the radiating patch. Moreover, pattern diversity characteristics and high inter-element isolation are also obtained by orthogonally placing the elements. In [21], a UWB MIMO antenna is proposed. The proposed structure is capable of rejecting 5.1–5.9 GHz WLAN band with high isolation of around –17 dB that has been obtained by introducing T-shaped slot. Similarly, T-shaped slot and line slot are used to achieve good impedance matching and high isolation [22]. Moreover, the band-notch characteristic is achieved by inserting two symmetrical C-shaped slots in the ground plane. A compact UWB MIMO antenna is proposed in [23]. The antenna consists of an inverted U-shaped slot to attain notch at the upper WLAN band whose performance matrix

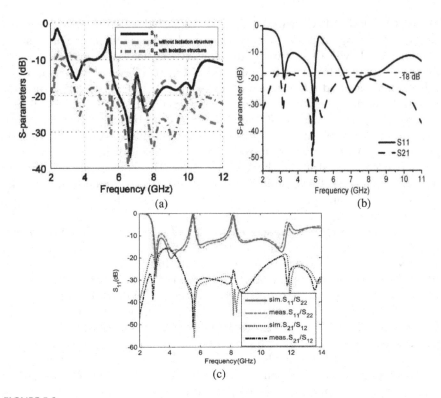

**FIGURE 5.2**
Simulated and measured results for slot loaded band-notched UWB MIMO antenna (a) $S_{11}$ and $S_{21}$ curve [23], (b) $S_{11}$ and $S_{21}$ curve [27], and (c) $S_{11}$ and $S_{21}$ curve [30].

is as shown in Figure 5.2(a). Moreover, the isolation between the elements is achieved by using modified ground plane. Two modified rectangular slot antennas excited by microstrip transmission lines are presented in [24]. To create band-notched characteristics at 5.15–5.825 GHz WLAN band, an inverted U-shaped slot is inserted in the feed line. Therefore, the mutual coupling reduction between elements is achieved by inserting T-shaped slot in the ground plane. Li et al. [25] have proposed a compact dual band-notched UWB-MIMO antenna with high isolation. The band-rejection functions for 5.15–5.85 GHz WLAN band and 3.3–3.7 GHz WiMAX band are realized by introducing one-third λ-rectangular metal strip and one-fourth λ-open slot into the radiator. Furthermore, two protruded ground parts are connected by a compact metal strip to reduce the mutual coupling between antenna elements. Quad band-notched UWB MIMO antenna is designed in [26]. Based on an inverted L-meander slot and two inverted L-shaped slots with C-shaped stubs near the feed line, quad notched bands are realized at 3.25–3.6 GHz, 5.05–5.48 GHz, 5.6–6 GHz, and 7.8–8.4 GHz, respectively. Furthermore, a T-shaped stub is realized over the ground plane for obtaining −20 dB isolation. A compact UWB MIMO antenna with band-notched

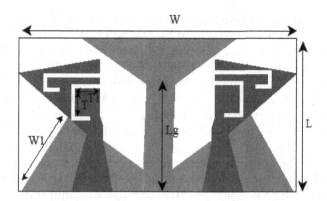

**FIGURE 5.3**
J-shaped slits for band-notch realization in a two-port MIMO antenna [28] [re-drawn].

characteristic is presented in [27]. A narrow line slot is etched into each element to obtain the notched band for blocking WLAN band. Next by introducing a decoupling slot into the ground plane, isolation between antennas with wider bandwidth is realized. Moreover, the proposed antenna has unsymmetrical dumbbell slot to obtain miniaturization. The performance parameters for the proposed design are shown in Figure 5.2(b). Kumar et al. [28] have designed a compact dual band-notched UWB MIMO antenna, as portrayed in Figure 5.3. The antenna consists of a triangular radiator shape with two J-shaped slits above the radiator to obtain notched-band functions from 5.1 GHz to 5.8 GHz and from 6.7 GHz to 7.1 GHz, respectively. Besides this, the diversity performance with low mutual coupling of –24 dB and envelope correlation coefficient (ECC) of less than 0.2 for the proposed antenna is also analyzed. Similarly, for achieving triple band-notched characteristics, C-shaped slot and S-shaped slot on the radiator are realized [29]. Furthermore, a T-shaped stub is applied on the ground plane to reduce mutual coupling below –20 dB. Zehnya et al. [30] have designed UWB MIMO Vivaldi antenna with dual band-notched characteristics at WLAN and X-band. The notches are realized by adding two SRRs of different sizes next to the microstrip feed lines. Moreover, by utilizing a T-shaped slot on the ground plane, isolation of –16 dB has been obtained and is shown in Figure 5.2(c). Triple band-notched UWB MIMO antenna is designed by embedding different types of slots and slits on the square patch [31]. The notches are obtained at 3.3–3.7 GHz WiMAX, 5.15–5.875 GHz WLAN, and 7.1–7.9 GHz X-band, respectively. The proposed antenna realizes low mutual coupling with lower ECC (<0.02) by introducing four-directional staircase-shaped decoupling, and multi-slit and multi-slot techniques. In addition to this, the suggested antenna produces stable gain, and quasi-omni-directional radiation patterns at the entire impedance bandwidth. Dual notched-band UWB MIMO antenna consisting of CSRR and SRR is proposed in [32]. The suggested geometry achieves an isolation of –15 dB within the specified UWB region by incorporating SRR structure to the ground plane. Furthermore, the notched-band functions are realized by placing CSRR to the radiating patch. Similarly, by inserting four slot-type SRRs on patch and ground of UWB MIMO antenna, single band rejection from 5.2 GHz to 5.9 GHz is achieved in [33]. Besides this, the proposed structure provides low mutual coupling of less than –20 dB. Gao et al. in [34] have proposed a MIMO/diversity antenna with WLAN notched band for UWB applications. The notched band at 5.5 GHz is achieved by etching two SRR slots on the radiator. To obtain high isolations, a rectangular stub is placed at 45 degrees between the radiators. Dual notched-band UWB MIMO antenna is designed using CSRR slots etched in both the radiators. The notches are obtained at 3.4–3.6 GHz WiMAX and 5.725–5.825 GHz WLAN with an isolation of –20 dB using ground stub and slots. Overall, the suggested design shows good performance in terms of impedance matching and stable radiation pattern [35]. Two configurations of quadruple band-notched UWB MIMO antenna with polarization diversity are proposed in [36]. Utilization of symmetric SRRs and CSRR near the feed line and radiating patch results in the realization of quad band-notched characteristics. In addition to this, the suggested antenna provides an isolation of –20 dB with increase in impedance bandwidth. Triple notched-band UWB MIMO antenna with high isolation is proposed by Huang et al. [37]. The antenna consists of four parasitic C-shaped SRRs (PCSRRs) and an inverted U-shaped slot for realization of notched bands covering 5.15–5.35 GHz, 5.725–5.825 GHz, and 3.7–4.2 GHz. In [38], two half-wavelength SRR slots are etched on the radiators of UWB MIMO antenna for creating notched-band functions within 5.15–5.9 GHz. Besides this, the suggested antenna provides an isolation of –15 dB by placing one rectangle stub placed at 45 degrees axis between the two CPW feed systems. A comparison in Table 5.1 has been made with different studies.

## 5.3 Parasitic Elements/Stub-Loaded Geometries

For creating band-notched characteristics in UWB antennas, loading of parasitic elements or stubs is also a very common technique. Dual band-notch MIMO antenna with two U-shaped patches has been evolved for UWB applications, as shown in Figure 5.4(a). Thus, by introducing trekking slots into the patch and C-shaped strips loaded around the feed line, notched-band characteristics are generated within the WiMAX (3.3–3.7 GHz) and WLAN (5.2–5.8 GHz) bands [39]. Finally, to reduce the mutual coupling between elements, a fork-shaped stub is loaded into the ground plane. A single band-notched UWB MIMO antenna is presented in [40]. The antenna comprises two radiators separated by an isolated element to reduce the mutual coupling between them. For creating the notched function, parasitic slits are placed on both the elements. Moreover, the design has good performance characteristics in terms of improving gain and radiation patterns with ECC less than about 0.07. Dual polarized band-rejection UWB MIMO antenna is

**TABLE 5.1**

Slot-Loaded Band-Notched UWB MIMO Antennas: A Comparison

| Ref. | Design Approach | Notched Band (GHz) | Isolation (dB) | Overall Geometry (mm³) |
|---|---|---|---|---|
| [11] | I/C shaped slot | (3.3–3.8)/(5.15–5.35) | –20 | 58×56×0.8 (= 2598) |
| [12] | Open-ended slot | (3.3–3.7)/(5.15–5.825) | –17 | 35×23×1.6 (= 1288) |
| [14] | U-shaped slot | 3.5/5.5 | –17 | 38×60×1.6 (= 3648) |
| [15] | U-shaped slot | (5.15–5.825)/(7.25–7.75) | –20 | 22×36×1.6 (= 1267) |
| [18] | L-shaped slot | (3.21–3.98)/(5.4–5.9) | –19.7 | 26.5×30×1.6 (= 1272) |
| [20] | Slit-cuts | (5.15–5.825) | –20 | 36×36×1.6 (= 2073) |
| [22] | C-shaped slot | (5.15–5.9) | –15 | 22×30×0.8 (= 528) |
| [23] | Inverted U-slot | (5–5.9) | –15 | 45.5×33×1.524 (= 2288) |
| [24] | Inverted U-slot | (5.1–5.825) | –20 | 22×26×0.8 (= 457) |
| [27] | Line slot | (5.1–6.0) | –18 | 24×20×0.8 (= 384) |
| [29] | C/S-shaped slot | 3.5/5.5/7.5 | –20 | 31×26×0.8 (= 644) |
| [30] | SRR | (5.3–5.8)/(7.85–8.55) | –16 | 26×26×0.762 (= 515) |
| [32] | CSRR/SRR | (3.3–3.6)/(4.8–5.0) | –15 | 54×33×0.8 (= 1425) |
| [35] | SRR | (3.4–3.6)/(5.725–5.825) | –20 | 35×30×0.8 (= 840) |
| [36] | CSRR/SRR | (3.4–3.8)/(4.8–5.6)/(5.7–6.4)/(7.7–8.6) | –20 | 21×46×1.5 (= 1449) |
| [37] | PCSRR | (3.7–4.2)/(5.15–5.35)/(5.725–5.825) | –21 | 30×26×0.8 (= 624) |

designed by Zhu et al. [41]. Two QSC antenna patches are being used for obtaining polarization diversity and better isolation (see Figure 5.4(b)). Notched band at WLAN system has been realized by etching a bent slit in each of the radiating elements. Further, the authors have modified the design structure with four-element MIMO system to observe the performance behavior.

Tang and Lin in [42] have proposed a new technique for achieving dual band-notched functions with good isolation in UWB MIMO antenna. The band rejection at 3.4–3.7 WiMAX, and at 5.15–5.35 GHz and 5.725–5.825 GHz WLAN bands is realized by introducing strips, which not only generate notches but also provide good isolation between elements. The performance parameter of the corresponding antenna is now shown in Figure 5.5(a). In [43], T-shaped stubs and L-shaped stubs are used for generating notched-band characteristics with good isolation between two ports, as depicted in Figure 5.4(c). The proposed antenna produces an impedance bandwidth ranging from 2.9 GHz to 20 GHz and having stable radiation patterns with ECC less than 0.012 and diversity gain (DG) more than 9.95. Similarly, T-shaped strip is used between antenna elements for reducing mutual coupling [44]. In addition to this, band rejection at 5.5 GHz is realized by etching a pair of L-shaped slits on the ground. Besides this, the antenna provides a low envelope correlation coefficient of less than 0.02. The performance parameter is shown in Figure 5.5(b). Four elements UWB MIMO antenna with dual notched-band characteristics have been presented in [45]. The reject bands at 3.28–4.3 GHz

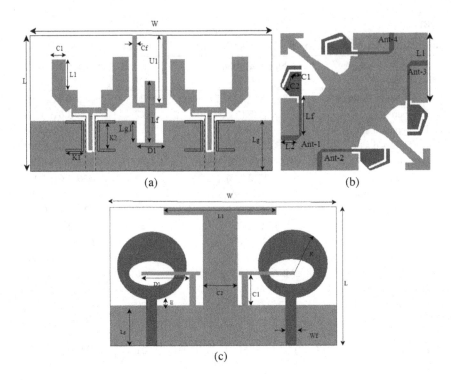

(a)                                        (b)

(c)

**FIGURE 5.4**
Parasitic elements/stubs-loaded band-notched UWB MIMO antennas [re-drawn] (a) C-shaped strips [39], (b) QSC antenna [41], and (c) T/L-shaped stubs [43].

and 4.9–5.5 GHz are achieved by using open circuit stubs of quarter wavelength in the ground plane. However, without any decoupling structure, the antenna provides isolation of around –20 dB. Liu et al. in [46] have designed a UWB MIMO antenna with single band-rejection characteristics. The antenna utilizes a T-shaped ground stub, and a vertical slot cut on the T-shaped ground stub to reduce mutual coupling below –15 dB. In addition to this two strips on the ground plane are being used to create a notched frequency band within the 5.15–5.85 GHz. Recently, a compact differential UWB MIMO antenna with single band-notched characteristics has been presented in [47]. The antenna consists of U-shaped microstrip feed line and a stepped-shaped slot with half-wavelength resonant stub for creating notched characteristics. Therefore, the proposed antenna provides an impedance of 2.95–10.8 GHz, and is having isolation better than 15.5 dB. The $S_{11}$ and $S_{21}$ curves are shown in Figure 5.5(c). A compact quad band-notched UWB MIMO antenna is proposed in [48]. Therefore, by inserting two L-shaped slots, CSRR and C-shaped stubs, four notched bands are achieved (3.25–3.9 GHz, 5.11–5.35 GHz, 5.5–6.06 GHz, and 7.18–7.88 GHz for WiMAX, lower WLAN, upper WLAN, and X-band, respectively). Furthermore, a comparison is provided in Table 5.2.

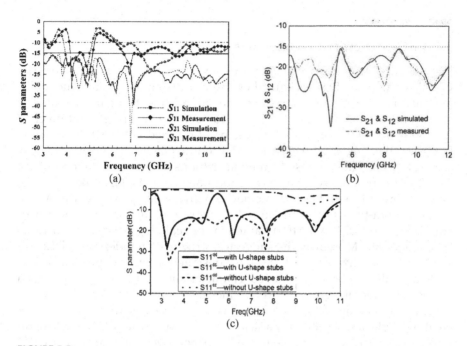

**FIGURE 5.5**
Simulated and measured results for parasitic elements/stubs-loaded band-notched UWB MIMO antenna (a) $S_{11}$ and $S_{21}$ curve [42], (b) $S_{11}$ and $S_{21}$ curve [44], and (c) $S_{11}$ and $S_{21}$ curve [47].

**TABLE 5.2**

Parasitic Element/Stub-Loaded Band-Notched UWB MIMO Antennas:
A Comparison

| Ref. | Design Approach | Notched Band (GHz) | Isolation (dB) | Overall Geometry (mm³) |
|------|----------------|--------------------|----------------|------------------------|
| [39] | C-shaped strips | (3.3–3.7)/(5.2–5.8) | – 20 | 60×35×1.6 (= 3360) |
| [40] | Parasitic slits | 5.8 | – 17 | 40.2×54×0.8 (= 1736) |
| [41] | Bent slits | (5.15–5.825) | – 20 | 35×35×1 (= 1225) |
| [42] | Parasitic strips | (3.4–3.7)/(5.15–5.35) | – 15 | 30×40×0.8 (= 960) |
| [43] | T/L-shaped stubs | (3.62–4.77) | – 20 | 18×36×1.6 (= 1036) |
| [44] | L-shaped slits | (5.03–5.97) | – 15 | 38.5×38.5×1.6 (= 2371) |
| [45] | Open circuit stubs | 3.3/5.0 | – 20 | 36×36×0.762 (= 987) |
| [46] | T-shaped stubs | (5.15–5.85) | – 15 | 22×36×1.6 (= 1267) |
| [47] | Resonant stubs | (5.1–5.95) | – 15 | 44×44×1.6 (= 3097) |
| [48] | C-shaped stubs | (3.25–3.9)/(5.11–5.35)/ (5.5–6.06)/(7.18–7.88) | – 18 | 45×30×0.8 (= 1080) |

## 5.4 EBG-Structured Geometries

Recently, EBG structures are being used for generating notch characteristics in UWB antennas. Besides this, they are also being used for reduction in mutual coupling in antenna arrays and improvement in gain. Literatures [49–52] have used various EBG structures for generating band-notched functions in UWB antennas. A single band-notched UWB MIMO antenna is designed using modified defected ground plane and a periodic EBG structure [49]. The proposed antenna exhibits impedance bandwidth of 3–16.2 GHz, and is having a sharp band-notched at 4.6 GHz. Besides this, an isolation of 17.5 dB has been achieved with a peak gain of 8.4 dB. A pair of mushroom-type EBG structures (See Figure 5.6(a)) has been used over circular monopole UWB MIMO antenna for realizing band-notched function at 5.8 GHz. Moreover, the suggested structure provides an isolation of 20 dB [50]. Mousazadeh et al. [51] have designed a compact printed MIMO antenna with triple band-notched characteristic by inserting two folded V-slot in the radiating patch and two modified mushroom-type EBG structures on either side of the feed line. Two folded V-slots are responsible for creating triple notched-band characteristics. However, modified mushroom-type EBG structures are implemented between two antennas. A 4×4 UWB MIMO antenna with band rejection at 5.5 GHz is shown in Figure 5.6(b). The notched-band function is obtained by implementing two mushroom-type EBG structures near the feedline. Moreover, an isolation of –15 dB is achieved by using rectangular stub. A comparison has been made for different EBG structures, as shown in Table 5.3.

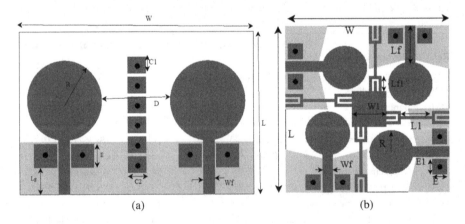

(a)                                      (b)

**FIGURE 5.6**
EBG-loaded band-notched UWB MIMO antenna [re-drawn] (a) mushroom-type EBG structure [50], and (b) modified mushroom-type EBG structures [52].

**TABLE 5.3**

EBG-Loaded Band-Notched UWB MIMO Antennas: A Comparison

| Ref. | Design Approach | Notched Band (GHz) | Isolation (dB) | Overall Geometry (mm³) |
|------|-----------------|--------------------|----------------|------------------------|
| 49 | Periodic EBG | 4.6 | −17.5 | 60×60×1.6 (= 5760) |
| 50 | Mushroom-type EBG | 5.8 | −20 | 30×90×0.813 (= 2195) |
| 51 | Modified mushroom-type EBG | 3.6/5.5/8.1 | −10 | – |
| 52 | Mushroom-type EBG | 5.5 | −15 | 30×30×1.6 (= 1440) |

# References

[1] O. Parchin et al., "Multi-band MIMO antenna design with user-impact investigation for 4G and 5G mobile terminals," Sensors, vol.19, p. 456, 2019.

[2] J. Ren, W. Hu, Y. Yin, R. Fan, "Compact printed MIMO antenna for UWB applications," IEEE Antennas and Wireless Propagation Letters, vol. 13, pp. 1517–1520, 2014.

[3] Y. Lee, D. Ga, J. Choi, "Design of a MIMO antenna with improved isolation using MNG metamaterial," International Journal of Antennas and Propagation, vol. 2012, Article ID 864306, p. 7, 2012.

[4] K. J. Babu, R. W. Aldhaheri, M. Y. Talha, I. S. Alruhaili, "Design of a compact two element MIMO antenna system with improved isolation," Progress In Electromagnetics Research Letters, vol. 48, pp. 27–32, 2014.

[5] FCC, Washington, DC, Federal Communications Commission revision of part 15 of the commission's rules regarding ultra-wideband transmission systems. First reported Order FCC: 02. V48, 2002.

[6] M. Rahman, S. D. Ko, D. J. Park., "A compact multiple notched ultra-wide band antenna with an analysis of the CSRR-TO-CSRR coupling for portable UWB applications," Sensors, Switzerland, vol. 17, 2017.

[7] A. Saxena, S. P. R. Gangwar, "Review on Band-Notching Techniques for Ultra Wideband Antenna." In: Nath V. (ed.) Proceedings of the International Conference on Nano-Electronics, Circuits & Communication Systems. Lecture Notes in Electrical Engineering, vol. 403, Singapore: Springer, 2017.

[8] B. Wang, H. Wang, H. Yu, "A very compact UWB-MIMO antenna with band-notched characteristic," IEEE International Workshop on Electromagnetics, Applications and Student Innovation Competition, Kowloon, pp. 124–127, 2013.

[9] S. Tripathi, A. Mohan, S. Yadav, "A Compact Octagonal Shaped Fractal UWB MIMO Antenna with 5.5 GHz Band-Notch Characteristics," In: IEEE International Microwave and RF Conference (IMaRC), Bangalore, pp. 178–181, 2014.

[10] Y. Liu, C. Sun, "A compact printed MIMO antenna for UWB application with WLAN band-rejected," International Symposium on Antennas, Propagation and EM Theory (ISAPE), Guilin, pp. 95–97, 2016.

[11] M. Lin, Z. Li, "A compact 4 × 4 dual band-notched UWB MIMO antenna with high isolation," IEEE International Symposium on Microwave, Antenna, Propagation, and EMC Technologies (MAPE), Shanghai, pp. 126–128, 2015.

[12] J. Li, D. Wu, Y. Wu, G. Zhang, "Dual Band-Notched UWB MIMO Antenna," In: IEEE Asia-Pacific Conference on Antennas and Propagation (APCAP), Kuta, pp. 25–26, 2015.

[13] K. Chhabilwad, G. S. Reddy, A. Kamma, B. Majumder, J. Mukherjee, "Compact dual band notched printed UWB MIMO antenna with pattern diversity," IEEE International Symposium on Antennas and Propagation & USNC/URSI National Radio Science Meeting, Vancouver, BC, pp. 2307–2308, 2015.

[14] S. Naser, N. Dib, "A Compact Printed UWB Pacman-Shaped MIMO Antenna with Two Frequency Rejection Bands," In: IEEE Jordan Conference on Applied Electrical Engineering and Computing Technologies (AEECT), Amman, pp. 1–6, 2015.

[15] A. Quddus, R. Saleem, S. Rehman, M. F. Shafique, "Dual Port UWB Diversity/ MIMO Antenna with Dual Band-Notch Characteristics," In: International Conference on Signal Processing and Communication Systems (ICSPCS), Gold Coast, QLD, pp. 1–4, 2016.

[16] M. Kapil, M. Sharma, "Dual Notched Band UWB Wireless MIMO Antenna," In: International Conference on Signal Processing and Integrated Networks (SPIN), Noida, India, pp. 8–12, 2019.

[17] K. Yu, Y. Kong, Y. Li, "A Two-Element UWB-MIMO Antenna with Quad Narrowband Frequency Rejection Characteristics," In: IEEE/ACES International Conference on Wireless Information Technology and Systems (ICWITS) and Applied Computational Electromagnetics (ACES), Honolulu, HI, pp. 1–2, 2016.

[18] Y. Li, Y. Kong, K. Yu, "A Dual Band-Notched UWB-MIMO Antenna Using Slot and Stub Techniques," In: IEEE Asia-Pacific Conf. on Antennas and Propagation (APCAP), Kaohsiung, pp. 313–314, 2016.

[19] Y. Yang, L. Jin, "Compact Dual-Polarized UWB-MIMO Antenna with Two Band-Notched Function," In: Proceedings of International Conference on Communication and Information Processing (ICCIP '16). Association for Computing Machinery, New York, NY, USA, pp. 206–210, 2016.

[20] J. Aquil, D. Sarkar, K. V. Srivastava, "A Quasi Self-Complementary UWB MIMO Antenna Having WLAN-Band Notched Characteristics," In: IEEE Applied Electromagnetics Conference (AEMC), Aurangabad, pp. 1–2, 2017.

[21] G. Irene, A. Rajesh, "Design of Orthogonal Dual Port UWB MIMO Antenna With IEEE 802.11ac Band Notch Characteristic," In: International Conference on Signal Processing, Communication and Networking (ICSCN), Chennai, pp. 1–5, 2017.

[22] Y. Zhang, X. Wu, Y. Li, Z. Liu, "A compact printed UWB MIMO antenna with WLAN band rejection," Progress in Electromagnetic Research Symposium (PIERS), Shanghai, pp. 2464–2466, 2016.

[23] T. Asghar, B. Ijaz, K. S. Alimgeer, M. S. Khan, R. Shubair, "A compact UWB MIMO antenna with inverted U-shaped slot for WLAN rejection," IEEE International Symposium on Personal, Indoor, and Mobile Radio Communications (PIMRC), Montreal, QC, pp. 1–4, 2017.

[24] F. Latif, F. A. Tahir, M. U. Khan, "Compact UWB-MIMO antenna with band-rejection in WLAN," Mediterranean Microwave Symposium (MMS), Abu Dhabi, pp. 1–4, 2016.

[25] J. Li, Q. Chu, Z. Li, X. Xia, "Compact dual band-notched UWB MIMO antenna with high isolation," IEEE Transactions on Antennas and Propagation, vol. 61, no. 9, pp. 4759–4766, 2013.

[26] L. Wu, Y. Xia, X. Cao, Z. Xu, "A miniaturized UWB-MIMO antenna with quadruple band-notched characteristics," International Journal of Microwave and Wireless Technologies, vol. 10, no. 8, pp. 948–955, 2018.

[27] Z. Yang, F. Li, F. Li, "A Compact Slot MIMO Antenna with Band-Notched Characteristic for UWB Application," In: International Conference on Microwave and Millimeter Wave Technology (ICMMT), Chengdu, pp. 1–3, 2018.

[28] A. K. Gautam, S. Yadav, K. Rambabu, "Design of ultra-compact UWB antenna with band-notched characteristics for MIMO applications," IET Microwaves, Antennas & Propagation, vol. 12, no. 12, pp. 1895–1900, 2018.

[29] Z. He, Z. Yang, J. Lv, "Design of a Novel Band-Notched Antenna for UWB MIMO Communication System," In: International Conference on Microwave and Millimeter Wave Technology (ICMMT), Chengdu, pp. 1–3, 2018.

[30] Z. Li et al. "Compact UWB MIMO Vivaldi antenna with dual band-notched characteristics." IEEE Access, vol.7, pp. 38696–38701, 2019.

[31] Z. Tang, X. Wu, J. Zhan, S. Hu, Z. Xi, Y. Liu, "Compact UWB-MIMO antenna with high isolation and triple band-notched characteristics," IEEE Access, vol. 7, p. 19856–19865, 2019.

[32] H. Li, Z. Jiang, "A CSRR and SRR based ultra wideband MIMO antenna with band-notched characteristics," IEEE International Symposium on Antennas and Propagation and USNC-URSI Radio Science Meeting, Atlanta, GA, USA, pp. 1137–1138, 2019.

[33] R. Xiao, X. Wei, L. Jin, "A Band-Notched UWB MIMO Antenna with High Notch-Band-Edge Selectivity," In: Asia-Pacific Microwave Conference (APMC), Nanjing, pp. 1–3, 2015.

[34] P. Gao, S. He, X. Wei, Z. Xu, N. Wang, Y. Zheng, "Compact printed UWB diversity slot antenna with 5.5-GHz band-notched characteristics," IEEE Antennas and Wireless Propagation Letters, vol. 13, pp. 376–379, 2014.

[35] Z. Wani, D. Kumar, "Dual-band-notched antenna for UWB MIMO applications," International Journal of Microwave and Wireless Technologies, vol. 9, no. 2, pp. 381–386, 2017.

[36] C. Xiaoyang, J. Long, "Compact Polarization Diversity Ultra-Wideband MIMO Antenna with Quadruple Band-Notched Characteristics," In: IEEE International Conference on Signal Processing (ICSP), Chengdu, pp. 1566–1570, 2016.

[37] H. F. Huang, B. Wang, "A Small-Size Ultra Wideband MIMO Antenna with Triple Band-Notched Function and High Isolation," In: IEEE International Conference on Computational Electromagnetics (ICCEM), Guangzhou, pp. 211–213, 2016.

[38] H. Shaung, G.Peng, "An UWB slot antenna with band-notched characteristics for MIMO/diversity applications," International Workshop on Microwave and Millimeter Wave Circuits and System Technology, pp. 138–141, 2013.

[39] D. D. Katre, R. P. Labade, "Higher isolated dual band notched UWB MIMO antenna with fork stub," IEEE Bombay Section Symposium (IBSS), Mumbai, pp. 1–5, 2015.

[40] K. Lin, L. Hwang, C. Hsu, S. Wang, F. Chang, "A Compact Printed UWB MIMO Antenna with A 5.8 GHz Band Notch," In: International Symposium on Antennas and Propagation Conference Proceedings, Kaohsiung, pp. 419–420, 2014.

[41] J. Zhu, S. Li, B. Feng, L. Deng, S. Yin, "Compact dual-polarized UWB quasi-self-complementary MIMO/diversity antenna with band-rejection capability," IEEE Antennas and Wireless Propagation Letters, vol. 15, pp. 905–908, 2016.

[42] T. Tang, K. Lin, "An ultra wideband MIMO antenna with dual band-notched function," IEEE Antennas and Wireless Propagation Letters, vol. 13, pp. 1076–1079, 2014.

[43] R. Chandel, A. K. Gautam, "Compact MIMO/diversity slot antenna for UWB applications with band-notched characteristic," Electronics Letters, vol. 52, no. 5, pp. 336–338, 2016.

[44] L. Kang, H. Li, X. Wang, X. Shi, "Compact offset microstrip-fed MIMO antenna for band-notched UWB applications," IEEE Antennas and Wireless Propagation Letters, vol. 14, pp. 1754–1757, 2015.

[45] D. Sipal, M. P. Abegaonkar, S. K. Koul, "Compact Planar Four Element Dual Band-Notched UWB MIMO Antenna for Personal Area Network Applications," In: European Conference on Antennas and Propagation (EuCAP), London, 2018, pp. 1–5.

[46] L. Liu, S. W. Cheung, T. I. Yuk, "Compact MIMO antenna for portable UWB applications with band-notched characteristic," IEEE Transactions on Antennas and Propagation, vol. 63, no. 5, pp. 1917–1924, 2015.

[47] Y. Liu, Z. Tu, "Compact differential band-notched stepped-slot UWB-MIMO antenna with common-mode suppression," IEEE Antennas and Wireless Propagation Letters, vol. 16, pp. 593–596, 2017.

[48] L. Wu, Y. Xia, "Compact UWB–MIMO antenna with quad-band-notched characteristic," International Journal of Microwave and Wireless Technologies, vol. 9, no. 5, pp. 1147–1153, 2017.

[49] W. Wu, B. Yuan, A. Wu, "A quad-element UWB-MIMO antenna with band-notch and reduced mutual coupling based on EBG structures," International Journal of Antennas and Propagation, vol. 2018, Article ID 8490740, p. 10, 2018.

[50] D. Z. Nazif, R. S. Rabie, M. A. Abdalla, "Mutual coupling reduction in two elements UWB notch antenna system," IEEE International Symposium on Antennas and Propagation & USNC/URSI National Radio Science Meeting, San Diego, CA, pp. 1887–1888, 2017.

[51] A. Mousazadeh, G. H. Dadashzadeh, "A novel compact UWB monopole antenna with triple band-notched characteristics with EBG structure and two folded V-slot for MIMO/diversity applications," ACES Journal, vol. 31, no. 1, 2016.

[52] N. K. Kiem, H. N. B. Phuong, D. N. Chien, "Design of compact 4×4 UWB-MIMO antenna with WLAN band rejection," International Journal of Antennas and Propagation, vol. 2014, Article ID 539094, p. 11, 2014.

# 6

## *Reconfigurable Band-Notched UWB Antennas*

## 6.1 Introduction

The rapid development of wireless communication has constituted the need for UWB systems with simple design configurations. According to the Federal Communications Commission (FCC) rulings, the allocated bandwidth for UWB system is 3.1–10.6 GHz. As a pivotal component of UWB systems, the antenna plays a vital role [1, 2]. Among the proposed UWB antennas, the printed monopole antenna (PMA) is very promising due to its remarkably small size, simple fabrication, and easy integration with compact RF front-ends [3, 4]. Nevertheless, underlying challenges of band overlapping in UWB antennas become serious problems. Some examples of the overlapping frequency bands are 3.3–3.6 GHz WiMAX, 3.7–4.2 GHz C-band, 5.15–5.35 GHz, and 5.725–5.825 GHz WLAN bands, which already exist within the UWB frequency band of 3.1–10.6 GHz [5, 6]. Therefore, to reject this interfering band, different techniques such as slotting of the radiator, adding parasitic elements near the radiator, and nowadays, EBG structures are also being used for generation of notched-band characteristics [7]. However, it has become necessary to have an antenna which can be operated both as a UWB antenna as well as a band-notched antenna, based on system requirements, for interference rejection as well as efficient spectrum utilization. To cater this demand, antennas with reconfigurable properties are highly demanded.

## 6.2 Slotted Geometries

Literature [4, 8–33] have used various slotted structures to attain band-notched characteristics. The reconfigurable characteristic is the introduced by employing different types of RF switches such as PIN diodes, Varactor diodes, MEMS switches and others. A novel UWB antenna with reconfigurable notch band is presented in [4]. The notched-band characteristic at 3.3–3.6 GHz is obtained by using a ring-shaped slot on the patch. To obtain a reconfigurable notch band, the ring-shaped slot structure is loaded with three switches. Two CPW-fed elliptical monopole antennas having reconfigurable band-rejection characteristics are reported in [8]. The antennas consist of a U-shaped slot and inverted L-shaped stubs to obtain the notched-band functions. Reconfigurability of the notched band (5.15–5.825 GHz) is produced by using micro electro mechanical system (MEMS) switches. Figure 6.1(a) reveals dual band-notched reconfigurable printed monopole antenna (RPMA), which is presented in [9]. By inserting two circular ring slots in the radiating patch, the antenna produces notches at 3.47–4.23 GHz and 5.11–5.94 GHz, which are further made switchable by inserting two PIN diodes within the slots. Boudaghi et al. in [10] have proposed a compact reconfigurable monopole antenna with two switchable band-notched characteristics at the frequencies 3–4 GHz and 5–6 GHz, respectively. Therefore, by using a U-shaped slot and a U-shaped parasitic patch in the back plane, the suggested antenna has realized dual notched characteristics. Moreover, the notched frequencies have been switched to other frequencies within the UWB range using PIN diodes. Reconfigurable UWB antenna with switchable band-notched characteristics for three different modes is presented in [11]. The antenna consists of a slotted circular ring with a rectangular patch to obtain notched characteristics (see Figure 6.1(b)). Based on three PIN diodes placed over the slotted circular ring, the notched switchable properties are obtained. A printed UWB antenna with switchable notched characteristics is presented in [12]. The antenna comprises a trapezoid-shaped patch with staircase slot in the ground plane for better matching. The rejection of interference at 3.5 GHz is realized by etching a complementary split-ring resonator on the radiating patch. Furthermore, the notch-band frequency is controlled by using PIN diodes. UWB antenna with switchable single, dual, and triple band-notch functions is presented in [13]. By etching arc-H-slot on the patch, ground, and feed line, (see

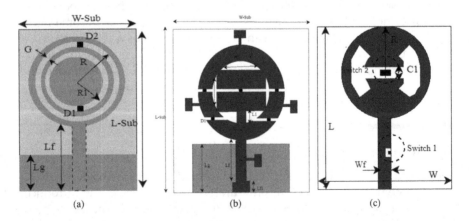

(a)                    (b)                    (c)

**FIGURE 6.1**
Different types of slot-loaded reconfigurable band-notched UWB antennas [re-drawn] (a) Circular ring slots [9], (b) Slotted circular ring [11], and (c) Arc-H-slot [13].

Figure 6.1(c)) the proposed antenna exhibits triple notched-band character-istics. Besides this, the antenna offers eight working modes of operation by turning appropriate switches on the circuit as ON or OFF. Similarly, by loading a folded slot on the U-shaped patch, a notched band of 2.2–2.9 GHz is obtained [14]. Thereafter, by using five laser controlled photoconductive silicon chips in the folded slot, reconfigurable band-notched characteristics at 2.4 GHz WLAN (2.2–2.9 GHz), WiMAX (3.2–4.7 GHz), 5.8 GHz WLAN (4.8–6.6 GHz), and ITU (7.5–8.7 GHz) are achieved.

A reconfigurable UWB monopole antenna with a switchable notched band is reported in [15]. The antenna uses a U-shaped slot, an open-ended quarter-wavelength slot, and two rectangular split loop resonators to achieve triple notched-band functions. The notched bands are made switchable with eight states of operation by adding three microstrip patches acting as switches instead of using traditional switches (Varactor diodes, PIN diodes). Recently, C-shaped slot and inverted T-shaped stubs have been introduced over the radiating patch to obtain dual notched-band characteristics within the range of 3.30 GHz–3.80 GHz (WiMAX) and 5.150 GHz–5.825 GHz (WLAN) [16]. In addition to this, PIN diodes are being used to reconfigure the proposed antenna with combination of different states of switches. Two reconfigurable band-notched monopole UWB antennas are presented in [17]. The antennas consist of a modified rectangular patch loaded with circular and rectangu-lar slots for proposing dual band-notched characteristics. Next, in order to achieve switchable characteristics, two PIN diodes are mounted over the circular slot, which results in single/dual/no notches, based on the switch-ing configurations. Kumar et al. [18] have designed a circular disc mono-pole UWB antenna with switchable dual notched bands. By cutting two U-shaped slots with folded arms on the radiating patch, dual band rejections at 3.25–3.83 GHz and 4.94–5.98 GHz are achieved. Finally, by introducing two switches on two slots, switchable notched frequency has been realized. Compact planar reconfigurable triple band-notched UWB antenna is pro-posed in [19]. Triple band-notched rejections are accomplished by inserting two slots with an inverted U-shaped metallic strip. The notches are realized at 3.6 GHz, 5.5 GHz, and 8.0 GHz, respectively. Furthermore, the proposed design provides single/dual band-notched behavior by embedding two PIN diodes along the patch slots, and the switchable operations for [18] and [19] are shown in Figure 6.2(a) and (b). Switchable band-notched UWB antenna with notch frequency at 5.8 GHz is proposed in [20]. The notched-band functions are realized by introducing two SRR structures back to the CPW feed. Thus, for introducing switchable behavior, two PIN diodes acting as a switch are mounted between the CPW conductor and the feed line, and the notch frequency is observed at 5.8 GHz when the switch is in OFF condi-tion. Meanwhile, the antenna acts as a narrowband antenna, with a center frequency of 5.8 GHz when the switch is ON (see Figure 6.2(c)). Luo et al. in [21] have proposed a compact UWB antenna with reconfigurable notch function. The antenna uses three different types of slots for generating three

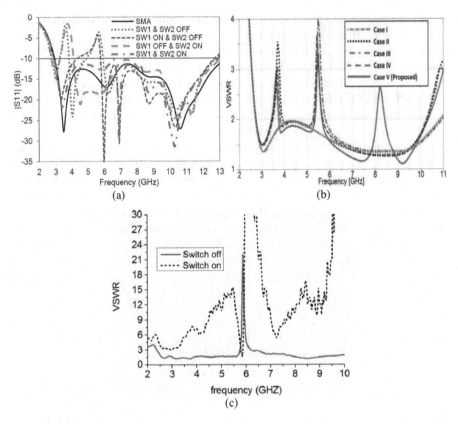

**FIGURE 6.2**
Simulated and measured results for slot-loaded reconfigurable band-notched UWB antenna (a) Simulated $S_{11}$ curve under different modes [18], (b) Simulated VSWR curve [19], and (c) Simulated VSWR curve under ON/OFF condition [20].

notched bands with central frequencies of 3.5, 5.5, and 7.5 GHz, respectively. Furthermore, to achieve reconfigurable function, three switches utilized across the slots are used to activate and deactivate the band-notched structures. In [22], a miniature UWB monopole antenna is studied. Hence, to achieve band-rejection behavior within 5–6 GHz, a U-shaped slot is embedded in the radiating patch. The performance of the proposed antenna is electronically controlled by a PIN diode located at the slot.

Switchable band-notched and multi-resonance UWB antenna is proposed by Valizade et al [23]. To create single band-notched characteristics within 5–6 GHz, a π-shaped slot is etched on the radiating stub. Furthermore, to achieve a reconfigurable function, a PIN diode is utilized across the slot, which transforms π-shaped slot into a pair of C-shaped slots, and results in switching the notched band from 5.03 GHz to 5.94 GHz. A UWB antenna with switchable notch bands of 3.25–4.26 GHz, 5.1–5.9 GHz, or 7.1–7.8 GHz is presented in [24]. The notched bands are realized by introducing a rectangular

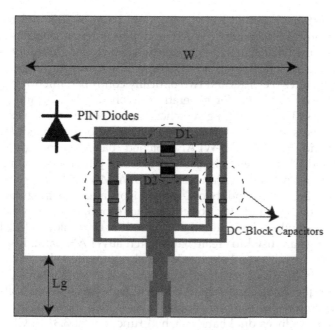

**FIGURE 6.3**
Inverted U-slots embedded with PIN diodes for reconfigurable band-notch characteristics [25] [re-drawn].

slot followed by parasitic patches and backplane structure. Three PIN diodes are placed at three different positions within the geometry for achieving the switching performance by adjusting the status of the PIN diodes. Switchable single and dual band-notch functions for UWB applications are presented in [25]. The notches are realized by embedding two symmetrical inverted U-slots on the radiating patch. Besides this, the antenna provides switchable notched-band properties by inserting two PIN diodes along these slots, which is shown in Figure 6.3. Cylindrical and conical UWB dielectric resonator antenna is presented in [26]. In order to minimize the interference between UWB and narrowband systems, a rectangular ring slot is etched on the radiating patch. However, by placing dielectric resonator at different angles, reconfigurability properties have been achieved from 3.2 GHz to 5.2 GHz. A reconfigurable dual band-notched U-shaped UWB antenna is presented in [27]. To create notched-band characteristics at WLAN and WiMAX bands, a T-shaped resonator and a complementary split ring resonator CSRR have been utilized. Thus, in order to produce reconfigurable band-notch characteristics, two switches are located over the T-shaped stub and CSRR. A novel butterfly-shaped UWB antenna with reconfigurable band notches is presented in [28]. The notched bands at 3.5 GHz and 5.6 GHz are realized by incorporating two pairs of circular split ring (CSR) of different sizes close to the feed line. Finally, by using four switches over the CSR, four switchable configurations have been attained. Similarly, by using dual pairs of SRR and rectangular spiral resonator (RSR) on the CPW-fed monopole antenna, dual band rejection has been obtained [29]. Furthermore, the notched reconfigurability is attained by proposing switches along the SRR and RSR based on three configurations. Likewise, two CSRRs and two optically controlled microwave switches (OCMSs) have been used for generating notched behavior and switchable characteristics [30]. On utilizing switches, the proposed antenna exhibits four reconfigurable states: (1) UWB only, (2) WiMAX-notched band, (3) WLAN-notched band, (4) WiMAX-WLAN-notched band. Reconfigurable triple notched-band UWB antenna is also presented in [31]. In order to reject triple notched frequency bands within 3.2–3.6 GHz WiMAX, 5.15–5.85 GHz WLAN, and 7.25–8.395 GHz X-band, two C-shaped slots and a rectangular slot have been loaded over the radiator. To achieve the reconfiguration in the notched-band antenna, authors have added a PIN diode at every slot. In [32], two identical split rings are used to create band-notch at WLAN band. In addition to this, the authors have used electronic switches mounted along these resonators to activate or deactivate the bands. Ammar et al. [33] have presented a novel technique for designing a UWB filtering antenna with dual sharp band-notches. Therefore, by using two double SRRs loaded above the ground plane, the antenna produces dual band-notched functions at 3.3–3.7 GHz WiMAX and 5.4–5.7 GHz HiperLAN2 frequencies. Furthermore, the reconfigurability feature is achieved by using two PIN diode switches employed to the two dual split ring resonators (DSRRs). Table 6.1 shows comparison between different slot loaded reconfigurable band-notched UWB antennas.

**TABLE 6.1**

Slot-Loaded Reconfigurable Band-Notched UWB Antennas: A Comparison

| Ref. | Design Approach | Notched-Band (GHz) | Overall Geometry (mm³) |
|------|-----------------|--------------------|------------------------|
| [9] | Circular ring slots | (3.47–4.23)/(5.11–5.94) | 18×12×0.8 (= 172) |
| [10] | U-slot | 3.5/5.5 | 22×22×1 (= 484) |
| [12] | Staircase slots | 3.5 | 22×32×1.6 (= 1126) |
| [13] | Arc-H-slot | (3.3–4.6)/(5.1–6.4)/(7.2–8.5) | 30×35.5×1.6 (= 1704) |
| [14] | Folded slot | (2.2–2.9)/(3.2–4.7)/(4.8–6.6)/(7.5–8.7) | 25×25×0.8 (= 500) |
| [15] | Quarter wavelength slot | 3.5/5.5 | 24×35×0.787 (= 661) |
| [16] | C-slot | (3.3–3.8)/(5.15–5.825) | 18×21×0.787 (= 297) |
| [17] | Rectangular slot | (3.12–3.84)/(5–6) | 20×20×0.8 (= 320) |
| [18] | U-slot | (3.25–3.83)/(4.94–5.98) | 30×40×1.6 (= 1920) |
| [20] | SRR | 5.8 | 50×50×1.574 (= 3935) |
| [22] | U-slot | 5–6 | 16×18×1.5 (= 432) |
| [24] | Ring slot | (3.25–4.26)/(5.1–5.9) | 17×14×0.8 (= 190) |
| [25] | Inverted U-slot | (3.15–3.85)/(5.43–6.1) | 20×20×0.8 (= 320) |
| [28] | SRR | 3.5/5.6 | 32×24×1.6 (= 1228) |
| [29] | SRR/RSR | 2.4/4.4 | 50×50×1.575 (= 3937) |

## 6.3 Parasitic-Element/Stub-Based Geometries

Wide impedance bandwidth with filtering characteristics in UWB antennas is the most challenging task. Therefore, the easiest way to accomplish this is by adding stubs and parasitic elements to the radiating patch and ground plane. A compact reconfigurable asymmetric coplanar strip (ACS)-fed monopole UWB antenna is presented in [34]. The proposed antenna rejects the interfering bands within the range of 3.5–4 GHz and 5.15–5.825 GHz by adding folded shape stub and slit to the structure. Furthermore, the band-reject reconfigurable ability is achieved via two RF PIN diode switches that are positioned in the slit. Similarly, by using folded shape slit and stub, dual notched-band characteristics of 3.3–4.2 GHz and 5–6 GHz have been accomplished [35]. The reconfigurable ability for the suggested geometry is achieved via a circular rotational motion of the patch. Circular ring resonator UWB antenna is presented by Kalteh et al. [36]. The antenna consists of a circular parasitic strip located above the circular ring with a circular slot etched on the ground plane. The extra arrangement over the substrate layer leads to the generation of notches. Besides this, the antenna reconfigurable capability over notch frequencies is accomplished by adding three PIN diodes between the T-shaped stub and the circular ring. Similarly, by inserting two inverted L-shaped stubs, band rejection at WLAN band is obtained [37]. The frequency notch can be potentially reconfigurable with the use of switches between stubs and elliptical radiator. Dual band-notched UWB antenna loaded with T-shaped stepped impedance resonator (T-SIR) and parallel stub loaded resonator (PSLR) is presented in Figure 6.4(a). The notches are

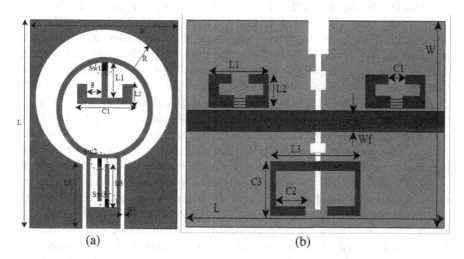

(a)                              (b)

**FIGURE 6.4**
Different types of parasitic elements/stubs-loaded reconfigurable band-notched UWB antennas [re-drawn] (a) T-SIR/PSLR [38], and (b) C-shaped resonators [50].

realized at 3.8–5.9 GHz and 7.7–9.2 GHz. The notch frequency reconfigurable characteristics are also obtained by integrating three switches into the T-SIR and the PSLR. It is also observed that the proposed antenna provides omni-directional radiation patterns with stable gain [38].

In [39], dual tunable and reconfigurable notch bands UWB antenna is reported. The dual band-notch function is implemented by using a stepped impedance resonator and a spiral-shaped stub. Meanwhile, the reconfigurable characteristics are realized by integrating two switches into the spiral-shaped stub and the stepped impedance resonator. Jacob et al. [40] have proposed a single band-notched UWB antenna. The band-notched function at WiMAX is obtained by etching S-shaped meandered parasitic element within the radiating aperture and can be switchable by placing ideal switches across the gaps of the parasitic element. Multimode reconfigurable dual band-notched UWB antenna is designed in [41]. The notched-band behavior is realized by integrating a two-stage T-shaped stepped impedance resonator (TSTSIR) inside the circular ring radiating patch, and by etching a parallel stubs loaded resonator (PSLR). In addition to this, the suggested geometry provides multimode reconfigurability based on the four ideal switches that are mounted into the TS-TSIR and the PSLR. Miswadi et al. [42] have designed filtenna with reconfigurable band-rejection for UWB applications. The proposed antenna comprises a circular radiation patch structure with partial ground plane, and has band-rejection properties. The notch at WLAN band is obtained by integrating straight open stub in the feed line. Furthermore, two modes of reconfigurability have been achieved on mounting RF switches over the open stub. Yadav et al. [43] have proposed frequency reconfigurable monopole antenna that has switchable notch characteristic at center frequency of 5.3 GHz. The notch at WLAN band is attained by integrating an F-shaped parasitic element with three stubs that are located on the back side of the radiating patch. Finally, the frequency reconfigurable properties are achieved by adding PIN diodes in between the ground plane and the F-shaped parasitic element. Wu et al. in [44] have proposed dual band-notched UWB antenna that has both switchable and tunable properties. Therefore, by using PIN diodes, the notch band can be switched to 4.2 GHz and 5.8 GHz, while the center notch frequency can be tuned from 4.2 GHz to 4.8 GHz and 5.8 GHz to 6.5 GHz, using varactor diode. Anti-interference antenna with excellent wide tunable and reconfigurable multiple filtering bands is proposed in [45]. The band-rejection functions are obtained by using three stubs into a CPW-fed circular slot. Besides this, the antenna exhibits tunable notched-band characteristics by adjusting the dimensions of stubs. Moreover, reconfigurable features are also being realized by integrating ideal switches into the stubs. Likewise, by inserting an inverted Γ-shaped parasitic element, single notched band at WLAN is obtained [46]. Furthermore, a rectangular split ring resonator is placed on the back side of the microstrip slot antenna to produce a notch at downlink X-band for satellite communication systems (7.2–7.8 GHz). Thus, in order to achieve reconfigurability in the proposed antenna, PIN diodes are being used under four different modes of operation. Hany et al. [47] have designed a CPW-fed UWB slot antenna with a single reconfigurable notched band. The proposed antenna utilizes two symmetrical short-circuited quarter wavelength resonators to create a single notch. By adjusting the length of resonators, switching characteristic over notched frequency has been attained. Moreover, the suggested antenna exhibits tunable behavior by inserting two ideal switches. Reconfigurable dual notched UWB antenna is designed in [48]. The antenna consists of a circular patch with two-pairs of L-shaped resonators in the radiating patch. The L-shaped resonators are responsible for generating notch characteristics in WLAN at 5.2 GHz and 5.8 GHz bands. In [49], authors have added a parasitic patch in the back plan to create band-notched in UWB antenna. The switchable notched properties are realized by cutting the parasitic element into two parts, which are then connected with two RF switches. Differential UWB antenna with reconfigurable band-notched characteristics using varactor diodes is proposed in [50]. By introducing three C-shaped resonators near the feed line (see Figure 6.4(b)), dual band rejection is achieved. The notches are realized within the frequency range of 4.5–5.3 GHz and 9.6–10.5 GHz. Finally, a varactor diode has been incorporated into the resonator to vary the notched frequencies. Similarly, by using a J-shaped stub inside a rectangular slot, single notched band at 5.6 GHz has been realized [51]. Therefore, to attain notched reconfigurability, a PIN diode has been hosted between the J-shaped stub and the monopole radiator. A comparison is made among the different parasitic elements/stub loaded reconfigurable band-notched UWB antennas which has been highlighted in Table 6.2.

**TABLE 6.2**

Parasitic Elements/Stubs-Loaded Reconfigurable Band-Notched UWB Antennas: A Comparison

| Ref. | Design Approach | Notched-Band/ Frequency (GHz) | Overall Geometry (mm³) |
|---|---|---|---|
| [34] | Folded-shaped stubs | (3.5–4)/(5.15–5.825) | 6.3×12×1 (= 172) |
| [36] | Parasitic strip/ T-stub | (3.5–4)/(5.15–5.825) | 45×40×1.6 (= 2880) |
| [37] | L-stub | 5.5 | 32×26×0.1 (= 83) |
| [38] | T-SIR/PSLR | (3.8–5.9)/(7.7–9.2) | 32×24×1.5 (= 1152) |
| [39] | Spiral stub | 3.5/5.5 | 32×24×1.6 (= 1228) |
| [41] | TS-TSIR/PSLR | 3.5/(7.7–8.5) | 24×32×1.6 (= 1228) |
| [42] | Open stub | (5.15–5.825) | 30×40×1.6 (= 1920) |
| [43] | F-shaped parasitic strip | 5.3 | 22×29×1.6 (= 1020) |
| [45] | U-stubs | 3.5/5.5/8.0 | 32×24×1.6 (= 1228) |
| [46] | Γ-shaped parasitic element | (5.1–5.7)/(7.2–7.8) | 24.8×30.3×0.8 (= 601) |
| [47] | Quarter wavelength resonator | (5.725–5.825)/(8.05–8.4) | 40×30×0.762 (= 914) |
| [50] | C-shaped resonator | 4.5–5.3 | 18×28×0.8 (= 403) |
| [51] | J-shaped stub | 5–6 | – |

## 6.4 EBG-Structured Geometries

To overcome the interfering issues, the EBG structures have been integrated with antennas/filters. Due to their inherent bandgap property, the EBG structures are being widely accepted among the researchers. The EBG structures are periodic collections of dielectric material and conductors. The concept of EBG structures originates from the solid state physics domain [52]. The bandgap feature of EBG has revealed the suppression of surface-wave in a particular frequency band, which results in improving the performance of the antenna. Another feature of the reflection phase of EBG structure can be intended to realize a perfect magnetic conductor (PMC)-like surface in a certain frequency band by using the reflection property. The reflection phase of an EBG surface varies continuously from +180° to − 180° [53]. Due to their unique bandgap property, EBG structures are widely used in several applications. A UWB antenna with reconfigurable band-notched characteristics using EBG structures is proposed in [54]. Two EBG structures are placed along the feed line of the UWB antenna to produce dual bands notched at 3.97 GHz and 5.51 GHz, which cover C-band satellite communication system and WLAN. Thus, in order to disable the EBG functions, a circular slot is introduced along with two switches. By switching the state of the switch, a different switchable notched band has been accomplished.

Majid et al. [55] have proposed a reconfigurable band-notched CPW-fed UWB antenna using EBG structure. The notched band at 4.0 GHz has been obtained by placing EBG structure adjacent to the transmission line. Furthermore, the band-notched characteristic can be disabled by switching the state of switch place at the stripline. Dual band-notched reconfigurable dual polarized UWB monopole antenna is presented in Figure 6.5(a). The notched band at 4.0–5.4 GHz is realized by placing a diagonally slotted tunable electromagnetic bandgap structure [56]. With properly adjusting the

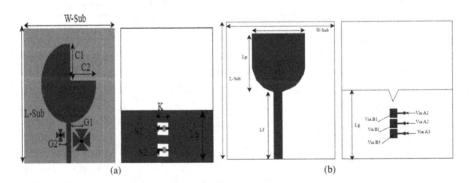

**FIGURE 6.5**
Simulated and measured results for slot loaded reconfigurable band-notched UWB antennas [re-drawn] (a) Diagonally slotted EBG [56], and (b) Mushroom-type EBG [57].

**TABLE 6.3**

EBG-Loaded Reconfigurable Band-Notched UWB Antennas: A Comparison

| Ref. | EBG Structure | Notched-Band/- Frequency (GHz) | Overall Geometry (mm³) |
|------|---------------|--------------------------------|------------------------|
| [54] | Mushroom-type EBG | 3.97/5.51 | 80×40×1.52 (= 4864) |
| [55] | Mushroom-type EBG | 4.0 | 50×80×1.6 (= 6400) |
| [56] | Diagonally slotted EBG | 8.2 | 67×67×1 (= 4489) |
| [57] | Mushroom-type EBG | 1.82 | 40×42×3.04 (= 5107) |

dimension of EBG structure, the notched band has been tuned. A wideband antenna with band-notch function using diagonally slotted EBG structure is proposed in [57]. Three EBGs are aligned underneath the feed line of the wideband antenna. Besides this, the proposed antenna produces reconfigurability properties by loading three switches; it is shown in Figure 6.5(b). Table 6.3 shows the comparison of different studies.

# References

[1] FCC, Washington, DC, Federal Communications Commission revision of part 15 of the commission's rules regarding ultra-wideband transmission systems. First reported Order FCC: 02. V48, 2002.

[2] P. P. Shome, T. Khan, R. H. Laskar, "A state- of- art review on band- notch characteristics in UWB antennas," International Journal of RF Microwave Computer-Aided Engineering, 2019.

[3] A. Sharma, M. Aggarwal, S. Ahuja, M. Uddin, "End-to-end performance of hybrid DF/AF (HDAF) relayed underlay cognitive radio networks," AEU-International Journal of Electronics and Communications, vol. 116, p. 153056.

[4] Y. Zhang, Q. Shi, S. Lin, S. Lu, "A novel reconfigurable notch-band UWB antenna," ISAPE, Xian, 2012, pp. 369–372.

[5] A. Saxena, R. P. S. Gangwar, "Review on Band-Notching Techniques for Ultra Wideband Antenna." In: Nath V. (ed.) Proceedings of the International Conference on Nano-Electronics, Circuits & Communication Systems. Lecture Notes in Electrical Engineering, vol. 403, Singapore: Springer, 2017.

[6] T. Saeidi, I. Ismail, W. P. Wen, A. R. H. Alhawari, A. Mohammadi, "Ultra-wideband antennas for wireless communication applications," International Journal of Antennas and Propagation, vol. 2019, Article ID 7918765, p. 25, 2019.

[7] T. Y. Yang, C. Y. Song, W. W. Lin, X. L. Yang, "A new band-notched UWB antenna based on EBG structure," International Workshop on Microwave and Millimeter Wave Circuits and System Technol, Chengdu, pp. 146–149, 2013.

[8] S. Nikolaou, N. D. Kingsley, G. E. Ponchak, J. Papapolymerou, M. M. Tentzeris, "UWB elliptical monopoles with a reconfigurable band notch using MEMS switches actuated without bias lines," IEEE Transactions on Antennas and Propagation, vol. 57, no. 8, pp. 2242–2251, 2009.

[9] N. Ojaroudi, S. Amiri, F. Geran, "Reconfigurable monopole antenna with controllable band-notched performance for UWB communications," Telecommunications Forum (TELFOR), Belgrade, pp. 1176–1178, 2012.

[10] H. Boudaghi, J. Pourahmadazar, S. A. Aghdam, "Compact UWB monopole antenna with reconfigurable band notches using PIN diode switches," WAMICON, Orlando, FL, pp. 1–4, 2013.

[11] D. Yadav, M. P. Abegaonkar, S. K. Koul, V. Tiwari, D. Bhatnagar, "Frequency Reconfigurable Monopole Antenna with Switchable Band Characteristics from UWB to Band-Notched UWB to Dual-Band Radiator," In: Asia-Pacific Microwave Conference (APMC), New Delhi, pp. 1–4, 2016.

[12] A. A. Mohammed, F. M. Alnahwi, A. S. Abdullah, A. G. A. A. Hameed, "A Compact Monopole Antenna with Reconfigurable Band Notch for Underlay Cognitive Radio Applications," In: International Conference on Advance of Sustainable Engineering and its Application (ICASEA), Wasit, pp. 25–30, 2018.

[13] S. Wang, J. Dong, M. Wang, "A Frequency-Reconfigurable UWB Antenna with Switchable Single/Dual/Triple Band Notch Functions," In: Cross Strait Quad-Regional Radio Science and Wireless Technology Conference (CSQRWC), Taiyuan, China, pp. 1–3, 2019.

[14] S. H. Zheng, X. Liu, M. M. Tentzeris, "Optically controlled reconfigurable band-notched UWB antenna for cognitive radio systems," Electronics Letters, vol. 50, no. 21, pp. 1502–1504, 2014.

[15] J. Li, Y. Sun, "Reconfigurable Triple Band-Notched Monopole UWB Antenna," In: Cross Strait Quad-Regional Radio Science and Wireless Technology Conference (CSQRWC), Taiyuan, China, pp. 1–3, 2019.

[16] M. Sharma, A. K. Goel, N. Kumar,Y. K. Awasthi, "Reconfigurable Dual Notched Band UWB Antenna," In: International Conference on Cloud Computing, Data Science & Engineering (Confluence), Noida, pp. 833–837, 2018.

[17] N. Tasouji, J. Nourinia, C. Ghobadi, F. Tofigh, "A novel printed UWB slot antenna with reconfigurable band-notch characteristics," IEEE Antennas and Wireless Propagation Letters, vol. 12, pp. 922–925, 2013.

[18] A. Kumar, I. B. Sharma, M. M. Sharma, "Reconfigurable circular disc monopole UWB antenna with switchable two notched stop bands," IEEE Annual India Conference (INDICON), Bangalore, pp. 1–4, 2016.

[19] W. A. E. Ali, R. M. A. Moniem, "Frequency reconfigurable triple band-notched ultra-wideband antenna with compact size," Progress In Electromagnetics Research C, vol. 73, pp. 37–46, 2017.

[20] K. Kandasamy, B. Majumder, J. Mukherjee, K. P. Ray, "Design of SRR loaded reconfigurable antenna for UWB and narrow band applications," IEEE International Symposium on Antennas and Propagation & USNC/URSI National Radio Science Meeting, Vancouver, BC, pp. 1192–1193, 2015.

[21] C. M. Luo, J. S. Hong, M. Amin, L. Zhong, "Compact UWB Antenna with Triple Notched Bands Reconfigurable," In: IEEE International Conference on Microwave and Millimeter Wave Technology, Beijing, pp.746–748, 2016.

[22] A. Vasylchenko, R. Dubrovka, W. D. Raedt, C. Parini, G. A. E. Vandenbosch, "Pulse Response Behavior of A UWB Antenna with Switchable Band-Notching Feature," In: European Conference on Antennas and Propagation (EUCAP), Rome, pp. 369–371, 2011.

[23] A. Valizade, C. Ghobadi, J. Nourinia, M. Ojaroudi, "A novel design of reconfigurable slot antenna with switchable band notch and multi-resonance functions for UWB applications," IEEE Antennas and Wireless Propagation Letters, vol. 11, pp. 1166–1169, 2012.

[24] S. Vahid et al., "A planar UWB antenna with switchable Single/Double band-rejection characteristics," Radioengineering, vol. 25, no. 3, 2016.

[25] B. Badamchi, J. Nourinia, C. Ghobadi, A. V. Shahmirzadi, "Design of compact reconfigurable ultra-wideband slot antenna with switchable single/dual band notch functions," IET Microwaves, Antennas & Propagation, vol. 8, no. 8, pp. 541–548, 2014.

[26] C. Aissaoui, I. Messaoudene, A. Benghalia, "Conical and Cylindrical DRAs with Reconfigurable Band Rejection for UWB Applications," In: International Multi-Conference on Systems, Signals & Devices (SSD), Marrakech, pp. 17–21, 2017.

[27] A. Kamma, G. S. Reddy, R. S. Parmar, J. Mukherjee, "Reconfigurable Dual-Band Notch UWB Antenna," In: National Conference on Communications (NCC), Kanpur, pp. 1–3, 2014.

[28] A. V. Golliwar, M. S. Narlawar, "Multiple Controllable Band Notch Butterfly Shaped Monopole UWB Antenna for Cognitive Radio," In: World Conference on Futuristic Trends in Research and Innovation for Social Welfare (Startup Conclave), Coimbatore, pp. 1–4, 2016.

[29] B. Belkadi, Z. Mahdjoub, M. L. Seddiki, M. Nedil, "UWB monopole antenna with reconfigurable notch bands based on metamaterials resonators," IEEE International Symposium on Antennas and Propagation & USNC/URSI National Radio Science Meeting, Boston, MA, pp. 285–286, 2018.

[30] D. Zhao, L. Lan, Y. Han, F. Liang, Q. Zhang, B. Wang, "Optically controlled reconfigurable band-notched UWB antenna for cognitive radio applications," IEEE Photonics Technology Letters, vol. 26, no. 21, pp. 2173–2176, 2014.

[31] B. Hammache, A. Messai, I. Messaoudene, M. A. Meriche, M. Belazzoug, F. Chetouah, "Reconfigurable Triple Notched-Band Ultra Wideband Antenna," In: International Conference on Innovations in Information Technology (IIT), Al-Ain, pp. 1–5, 2016.

[32] A. V. Golliwar, M. S. Narlawar, "Multiple Controllable Band Notch Antenna for UWB Cognitive Radio Application," In: International Conference on Signal Processing and Integrated Networks (SPIN), Noida, pp. 694–697, 2016.

[33] A. Alhegazi, Z. Zakaria, N. A. Shairi, I. M. Ibrahim, S. Ahmed, "A novel reconfigurable UWB filtering-antenna with dual sharp band notches using double split ring resonators," Progress In Electromagnetics Research C, vol. 79, pp. 185–198, 2017.

[34] P. Lotfi, M. Azarmanesh, E. A. Sani, S. Soltani, "Design of very small UWB monopole antenna with reconfigurable band-notch performance," International Symposium on Telecommunications (IST), Tehran, pp. 102–105, 2012.

[35] P. Lotfi, M. Azarmanesh, S. Soltani, "Rotatable dual band-notched UWB/triple-band WLAN reconfigurable antenna," IEEE Antennas and Wireless Propagation Letters, vol. 12, pp. 104–107, 2013.

[36] A. A. Kalteh, G. R. Dadashzadeh, M. N. Moghadasi, B. S. Virdee, "Ultra-wideband circular slot antenna with reconfigurable notch band function," IET Microwaves, Antennas & Propagation, vol. 6, pp. 108–112, 2012.

[37] S. Nikolaou, A. Amadjikpe, J. Papapolymerou, M. M. Tentzeris, "Compact Ultra Wideband Elliptical Monopole with Potentially Reconfigurable Band Rejection Characteristic," In: Asia-Pacific Microwave Conference, Bangkok, pp. 1–4, 2007.

[38] Y. Li, W. Li, Q. Ye, "A CPW-fed circular wide-slot UWB antenna with wide tunable and flexible reconfigurable dual notch bands," The Scientific World Journal, vol. 2013, ID 402914, p. 10, 2013.

[39] Y. Li, W. Yu, "Design of An Ultra Wideband Antenna with Tunable and Reconfigurable Band-Notched Characteristics," In: IEEE Asia-Pacific Conference on Antennas and Propagation (APCAP), Kuta, pp. 142–143, 2015.

[40] S. Jacob, S. Nimisha, P. V. Anila, P. Mohanan, "UWB Antenna with Reconfigurable Band-Notched Characteristics Using Ideal Switches," In: IEEE International Microwave and RF Conference (IMaRC), Bangalore, pp. 136–139, 2014.

[41] Y. Li, W. Li, Q. Ye, "A compact circular slot UWB antenna with multimode reconfigurable band-notched characteristics using resonator and switch techniques," Microwave and Optical Technology Letters, vol. 56, pp. 570–574, 2014.

[42] N. F. Miswadi, M. T. Ali, M. N. M. Tan, N. H. Baba, F. N. M. Redzwan, H. Jumaat, "A reconfigurable band-rejection filtenna using open stub for ultra wideband (UWB) applications," IEEE Symposium on Computer Applications & Industrial Electronics (ISCAIE), Langkawi, pp. 7–10, 2015.

[43] D. Yadav, M. P. Abegaonkar, S. K. Koul, V. N. Tiwari, D. Bhatnagar, "A novel frequency reconfigurable monopole antenna with switchable characteristics between band-notched UWB and WLAN applications," Progress In Electromagnetics Research C, vol. 77, pp. 145–153, 2017.

[44] W. Wu, Y. B. Li, R. Y. Wu, C. B. Shi, T. J. Cui, "Band-notched UWB antenna with switchable and tunable performance," International Journal of Antennas and Propagation, vol. 2016, Article ID 9612987, p. 6, 2016.

[45] Y. Li, X. Liu, K. Yu, Y. Wang, "A planar anti-interference UWB antenna with designated tunable and reconfigurable multiple filtering bands," Progress in Electromagnetic Research Symposium (PIERS), Shanghai, pp. 1977–1981, 2016.

[46] H. Oraizi, N. V. Shahmirzadi, "Frequency- and time-domain analysis of a novel UWB reconfigurable microstrip slot antenna with switchable notched bands," IET Microwaves, Antennas & Propagation, vol. 11, no. 8, pp. 1127–1132, 22 6 2017.

[47] H. A. Atallah, A. B. A. Rahman, K. Yoshitomi, R. K. Pokharel, "Reconfigurable band-notched slot antenna using short circuited quarter wavelength microstrip resonators," Progress In Electromagnetics Research C, vol. 68, pp. 119–127, 2016.

[48] S. W. Yik et al. "A compact design of reconfigurable dual band-notched UWB antenna," IEEE International Workshop on Electromagnetics: Applications and Student Innovation Competition (iWEM), pp. 1–2, 2018.

[49] A. H. Khidre, H. A. ElSadek, H. F. Ragai, "Reconfigurable UWB printed monopole antenna with band rejection covering IEEE 802.11a/h," IEEE Antennas and Propagation Society International Symposium, Charleston, pp. 1–4, 2009.

[50] N. Nie, Z. Tu, "Differential UWB Antenna with Reconfigurable Band-Notched Characteristics Using Varactor Diodes," In: International Conference on Microwave and Millimeter Wave Technology (ICMMT), Chengdu, pp. 1–3, 2018.

[51] A. Quddious, M. A. B. Abbasi, P. Vryonides, S. Nikolaou, M. A. Antoniades, B. Manhaval, "Reconfigurable notch-band UWB antenna with RF-to-DC rectifier for dynamic reconfigurability," IEEE Antennas and Propagation Society International Symposium and USNC/URSI National Radio Science Meeting, APSURSI, 2018.

[52] F. Yang, Y. R. Samii, Electromagnetic Band Gap Structures in Antenna Engineering, New York: Cambridge University Press, 2009.

[53] S. Raza, "Characterization of the reflection and dispersion properties of mushroom-related structures and their application to antennas," Thesis on Master of Applied Science, University of Toronto, 2012.

[54] H. A. Majid, M. K. A. Rahim, M. R. Hamid, N. A. Murad, A. Samsuri, O. Ayop, "Reconfigurable Band Notch UWB Antenna Using EBG Structure," In: IEEE Asia-Pacific Conference on Applied Electromagnetics (APACE), Johor Bahru, pp. 268–270, 2014.

[55] H. A. Majid, M. K. A. Rahim, M. R. Hamid, "Band-notched reconfigurable CPW-fed UWB antenna," Applied Physics, vol. 347, p. 122, 2016.

[56] K. Krishnamoorthy, B. Majumder, J. Mukherjee, "Dual Polarized Reconfigurable Dual Band-Notched UWB Antenna with Novel Tunable EBG Structure," In: IEEE Applied Electromagnetics Conference (AEMC), Bhubaneswar, pp. 1–2, 2013.

[57] H. A. Majid, M. K. Abd Rahim, M. R. Hamid, M. F. M. Yusoff, N. A. Murad, N. A. Samsuri, O. B. Ayop, R. Dewan, "Wideband antenna with reconfigurable band notched using EBG structure," Progress In Electromagnetics Research Letters, vol. 54, pp. 7–13, 2015.

# 7

# *Tunable Band-Notched*
# *UWB Antennas*

## 7.1 Introduction

Since the release of UWB technology having allocated a frequency band of 3.1–10.6 GHz by the FCC for commercial applications, it has gained huge attention because of its numerous advantages such as high data rates with low power consumability, and low cost fabrication [1]. To realize UWB communication system, UWB antennas with compact size and wider impedance bandwidth are highly in demand. As a result, various shapes of UWB antennas are designed; among these, printed monopole UWB antennas are very much popular due to their remarkably compact size, stable radiation characteristics, and ease of fabrication [2]. However, in practical applications, UWB systems suffer band overlapping challenges, i.e., some narrow-band frequencies such as WiMAX (3.3–3.7 GHz), C-band (3.77–4.2 GHz/5.9–6.4 GHz), WLAN (5–6 GHz), and X-band (8.025–8.4 GHz) already exist within the specified UWB frequency band [3, 4]. Therefore, to reject the interfering bands, UWB antennas with filtering properties are required. The most general methods for obtaining notched characteristics are by etching of slots in the radiating patch and ground plane, and by adding parasitic elements or stubs near the radiator [5]. However, the above mentioned techniques are used not only for obtaining notched characteristics but also for removing the requirements of additional bandstop filters. But unfortunately these designs permanently reject the notch-bands once it is fabricated. However, sometimes band rejection is not necessary when there is no co-existence with narrowband systems. To solve these problems, smart reconfigurable UWB antenna is required having the capability to reconfigure its notched behaviour as per the system requirement. Switching the band notch frequencies between different operating states has already been discussed in Chapter 6. However, an additional degree of adaptability can be achieved, when the notch frequencies are dynamically tuned. The tuning is performed using varactor diodes where the capacitance of the varactor is changed continuously by varying the reverse bias voltage applied across it. This variation in capacitance leads to alteration of resonance behavior of the antenna over a frequency spectrum [6].

## 7.2 Slotted Geometries

In this section, UWB antennas with tunable notched characteristics are discussed. Authors in [7–15] have designed various UWB antennas with slotted structures for achieving band-rejection properties where the tunable characteristics are achieved by utilizing MEMS/PIN/varactor diodes as switches. A UWB planar monopole antenna with a tunable band-notched response is proposed in [7]. The notched-band function at 5.8 GHz is accomplished by embedding a rectangular ring-shaped slot in the ground plane. Furthermore, band tuning functions of 4.8–6.2 GHz are also realized by loading a varactor diode on the rectangular slots. Xia et al. [8] have designed dual tunable band-notched printed monopole UWB antenna. The notched-band functions of 3.3–3.7 GHz for WiMAX and 5.15–5.825 GHz for WLAN bands are obtained by introducing C-shaped slots above and below the substrate layer, and their central notched frequencies are tuned individually by placing a capacitor and an inductor on each of the slots. A UWB antenna with single band-notched function at 5.8 GHz is presented in [9]. The antenna consists of a pyramidal radiating patch with slots in each of the faces of the pyramid. Therefore, by introducing lumped capacitor or varactor diode over the slots, the notch-band center frequency has been tuned from 4.8 GHz to 7.472 GHz by varying the capacitor value from 0.1 pF to 10 pF. A miniaturized UWB antenna with single tunable band-notched characteristic at 5.5 GHz is presented in [10]. The notched-band function is accomplished by realizing a C-shaped slot above the radiating patch.

The notched frequency tuning capability is achieved by varying the thickness of the slots from 0.1 mm to 1.4 mm at a step size of 0.05 mm. A novel UWB antenna with tunable notched-band function is presented by Sajjad [11]. By inserting a $\pi$-shaped slot (see Figure 7.1(a)) on the radiating patch, band-notch function is achieved. Thereafter, notched tunability property from 2.7 GHz to 7.2 GHz is achieved by introducing lumped varactor element over the slot. In [12], a UWB antenna with a compact coplanar waveguide resonant cell (CCRC) is proposed, and its switchable operations are shown in Figure 7.2(a). The suggested antenna produces a bandwidth of 3–10.3 GHz with a band-notch function in the range of 5.4–6.1 GHz. The center frequency of the created stopband is tuned by embedding varactor diodes in CCRC. Hence, it is observed that an increase in the varactor capacitance from 0.1 pF to 0.5 pF changes the center frequency of band-notch from 5.7 GHz to 3.1 GHz. Ali et al [13], have designed a UWB antenna using single layer of metamaterial structure. The suggested antenna consists of circular patch with stepped cuts and curvature for better impedance matching from 3.1 GHz to 12 GHz. The notched-band characteristics from 5.7 GHz to 5.9 GHz are realized by implementing $\Omega$-structure in the radiating patch. Further, lumped capacitor is inserted in omega structure to tune the notched frequency from 5.7 GHz WLAN band to 3.8 GHz WiMAX band.

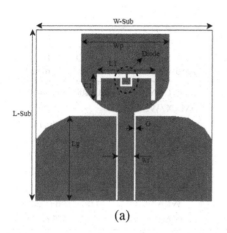
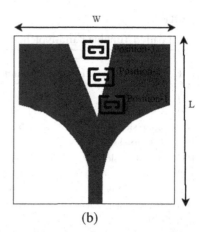

(a)                            (b)

**FIGURE 7.1**
Different types of slot loaded tunable band notched UWB antennas [re-drawn] (a) π-shaped slot antenna [11], and (b) Antipodal Vivaldi antenna [14].

Besides this, the antenna provides an improved gain by adding single array layer with inverted 3×2 L-shaped metamaterial unit cells. Sarkar et al. [14] have proposed an antipodal Vivaldi UWB antenna with tunable notched functions, as shown in Figure 7.1(b). The notched function within 5–6 GHz frequency range is achieved by placing a parasitic rectangular SRR near the radiating arm. Moreover, the notched center frequency is tuned by varying the position and dimensions of SRR structure (see Figure 7.2(b)). Similarly, triple band-notch Vivaldi antenna has been designed by Keerthipriya et al. [15]. The proposed antenna exhibits triple band-notch with tunable notch characteristics. The notched function is obtained by implementing triple SRR structures on the slot-line and tuning them using a varactor diode with a capacitance ranging from 0.1 pF (20 V) to 3 pF (0 V), and a parasitic capacitance of 0.06 pF. A comparison is made and shown in Table 7.1.

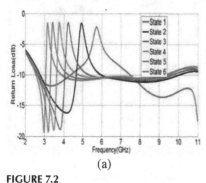
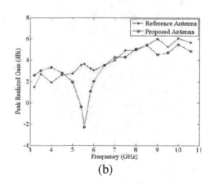

(a)                            (b)

**FIGURE 7.2**
Simulated and measured results for slot-loaded tunable band-notched UWB antenna (a) simulated $S_{11}$ curve [12], and (b) simulated gain curve [14].

**TABLE 7.1**

Slot-Loaded Tunable Band-Notched UWB Antennas

| Ref. | Design Approach | Notch-Band (GHz) | Notch-Band Applications | Overall Geometry (mm³) |
|------|-----------------|------------------|-------------------------|------------------------|
| [7] | Ring slot | 5.8 | WLAN | 100×100×0.76 (= 7600) |
| [8] | C-shaped slot | (3.3–3.7)/(5.15–5.825) | WiMAX/WLAN | – |
| [9] | Loop slots | 5.5 | WLAN | 80×80×2.33 (= 14912) |
| [10] | C-shaped slot | 5.5 | WLAN | 20×18×1.5 (= 540) |
| [11] | π-shaped slot | (5.8–7.9) | WLAN | 30×30×0.8 (= 720) |
| [12] | CCRC | (5.4–6.1) | WLAN | 30×30×0.8 (= 720) |
| [13] | Ω-shaped slot | (3.8–5.7) | WLAN | 35×35×1.575 (= 1929) |
| [14] | Rectangular SRR | (5–6) | WLAN | 11×32×1.6 (= 563) |
| [15] | SRR | 3.5 | WiMAX | 75×60×1.575 (= 7087) |

## 7.3 Parasitic-Elements/Stub-Loaded Geometries

For creating band-notched characteristics in UWB antennas, loading of parasitic elements or stubs is also a very common technique. However, these techniques make the design complex. In [16], tunable band-notched UWB planar monopole antenna is presented. The proposed antenna consists of a rectangular patch, quarter-wavelength stub loaded with a varactor diode. The notched-band characteristic at WLAN is obtained after implementing stub, which is further tuned by a variable capacitance diode (BBY52-02, Infineon Corp.) having a variance from 1.85 pF to 1.15 pF. Li et al. [17] have designed a compact UWB antenna with tunable notched frequency shown in Figure 7.3(a). The notched-band characteristic is obtained by using a pair of open-loop resonators (OLRs). The center frequency of the notched is further tuned

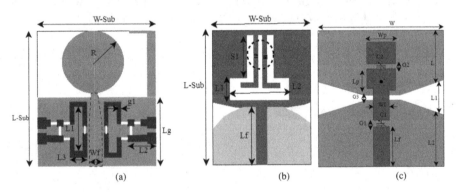

(a)　　　　　　　　(b)　　　　　　　　(c)

**FIGURE 7.3**

Different types of stub and parasitic element loaded tunable band notched UWB antennas [re-drawn] (a) Open-loop resonators antenna [17], (b) Stubs and strips loaded antenna [18], and (c) Shunt LC-resonator antenna [21].

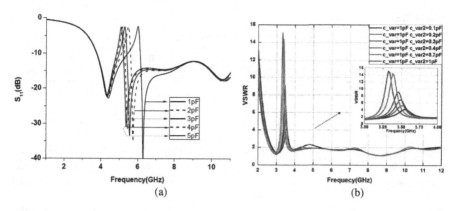

**FIGURE 7.4**
Simulated and measured results for stub and parasitic element loaded antennas (a) Simulated $S_{11}$ curve with variable capacitance [17], and (b) VSWR curve with variable capacitance [18].

using a varactor diode, which changes the effective electrical length of the OLRs. Besides this, the proposed antenna provides miniaturization using C-shaped ground. A novel UWB planar antenna with tunable band-notch characteristics is shown in Figure 7.3(b). The authors have used two strips with stubs and a varactor diode for realizing band-notched and tunable characteristics [18]. Therefore, by changing the bias voltage applied to the varactors, the band-notch center frequency is tuned from 3.32 GHz to 4.68 GHz. The performance characteristics for [17] and [18] are shown in Figure 7.4(a) and (b), respectively.

Multi band-notched UWB antenna having both tunable and reconfigurable properties is presented in [19]. The multiple anti-interference functions are obtained by using three stubs integrated into a CPW-fed circular wide slot. The notched tunability is realized by adjusting the dimensions of these stubs. Meanwhile, excellent reconfigurable features are obtained by integrating ideal switches. Similarly, a compact reconfigurable and tunable band-notched UWB antenna is proposed in [20]. The antenna consists of an S-shaped SRR for obtaining notched-band behavior. The notched tunability and reconfigurability are accomplished by loading S-SRR with a varactor diode. Therefore, by implementing a pair of shunt PIN diodes, notched frequency switching characteristics are attained. Based on varactor diode having capacitance range from 1.5 pF to 5.5 pF, a tunable notched UWB antenna is designed [21]. The proposed antenna with shorting pins and a shunt LC-resonator leads to the generation of notched characteristics. Therefore, the notched center is further tuned by using a varactor diode (see Figure 7.3(c)). Zhou et al. [22] have designed a compact UWB antenna with dual independent tunable band-notches. Thus, by utilizing a V-shaped strip and a pair of L-shaped slots on the radiation patch, dual band-notched functions at 3.2–3.8 GHz and 5–5.9 GHz are obtained. Moreover, the notched tunability properties are realized by varying the dimensions of V-shaped strips and L-shaped slots. A rectangular patch antenna consisting

**TABLE 7.2**

Parasitic Elements/Stubs-Loaded Tunable UWB Antenna Geometries

| Ref. | Design Approach | Notch-Band (GHz) | Notch-Band Applications | Overall Geometry (mm³) |
|---|---|---|---|---|
| [16] | Quarter wavelength stubs | (5.1–5.3) | WLAN | 88×70×0.76 (= 4681) |
| [17] | Open loop resonator | (5.1–5.9) | WLAN | 28×24×1 (= 672) |
| [18] | Strips/stubs | 3.3 | WiMAX | 27×34×0.787 (= 722) |
| [19] | Stubs | 5.5 | WLAN | 32×24×1.6 (= 1228) |
| [20] | S-shaped SRR | (3.3–3.6)/(5.15–5.35)/ (5.725–5.825) | WiMAX, WLAN | 49.4×35×0.257 (= 444) |
| [21] | LC resonator | (5–6) | WLAN | 28×28.5×0.8 (= 638) |
| [22] | V-shaped strip/ L-shaped slot | (3.2–3.8)/(5–5.9) | WiMAX, WLAN | 20×36×0.508 (= 365) |
| [23] | Open loop resonator | (5.1–5.6) | WLAN | 32×32×1.5 (= 1536) |
| [24] | Rectangular strip/ U-shaped strip | 5.0 | WLAN | 27×34×0.787 (= 722) |
| [25] | Strip resonator | (3–4) | WiMAX | 30×31×1.5 (= 1395) |

of stepped cuts and triangular slot is proposed in [23]. Two open loop resonators are designed and added near the microstrip feed line for achieving band rejection from 5.1 GHz to 6.5 GHz. Finally, by inserting two-lumped capacitors, the notched frequency tuning ability has been investigated. In [24], compact planar UWB antennas with two independent notched frequencies are investigated. With the help of a split straight rectangular strip and an interdigital U-shaped parasitic element above and beneath the substrate, the proposed antenna produces dual notched-band behavior, which is further tuned by inserting a pair of varactor diodes. A band-notched UWB antenna having continuously tunable WiMAX rejection band and fixed WLAN rejection band is presented in [25]. Therefore, by implementing a miniaturized resonator in the partial ground plane, notched tunability from 3 GHz to 4 GHz is achieved. Table 7.2 shows a comparison of different studies.

## 7.4 Summary

Due to the increasing demand for high data and low power consumption, researchers are looking for solutions that produce optimal results. UWB systems provide better solution for short-range communications with high data rate and low power consumption. In this chapter, UWB antennas with tunable notch-band characteristics are investigated and discussed. The realization of single/dual/triple notched-band functions using different techniques is reported. Further, the notched frequency tunability is accomplished using varactor diodes.

## References

[1] FCC, Washington, DC, Federal Communications commission revision of part 15 of the commission's rules regarding ultra-wideband transmission systems. First reported Order FCC: 02. V48, 2002.

[2] P. P. Shome, T. Khan, R. H. Laskar, "A state- of- art review on band- notch characteristics in UWB antennas," International Journal of RF and Microwave Computer Aided Engineering, 2019.

[3] M. Rahman, D. S. Ko, J. D. Park, "A compact multiple notched ultra-wide band antenna with an analysis of the CSRR-to-CSRR coupling for portable UWB applications," Sensors (Basel, Switzerland), vol. 17, 2017.

[4] A. Saxena, R. P. S. Gangwar, "Review on Band-Notching Techniques for Ultra Wideband Antenna," In: Proceedings of the International Conference on Nano-electronics, Circuits & Communication Systems. Lecture Notes in Electrical Engineering, vol. 403, Springer, Singapore, 2017.

[5] T. Saeidi, I. Ismail, W. P.Wen, A. R. H. Alhawari, A. Mohammadi, "Ultra-wideband antennas for wireless communication applications," International Journal of Antennas and Propagation, vol. 2019, Article ID 7918765, p. 25, 2019.

[6] R. Cicchetti, E. Miozzi, O. Testa, "Wideband and UWB antennas for wireless applications: A comprehensive review," International Journal of Antennas and Propagation, vol. 2017, Article ID 2390808, p. 45, 2017.

[7] E. A. Daviu, M. C. Fabres, M. F. Bataller, A. V. Jimenez, "Active UWB antenna with tunable band-notched behavior," Electronics Letters, vol. 43, no. 18, pp. 959–960, 2007.

[8] Y. Q. Xia, J. Luo, D. J. Edwards, "Novel miniature printed monopole antenna with dual tunable band-notched characteristics for UWB applications," Journal of Electromagnetic Waves and Applications, pp.1783–1793, 2010.

[9] Z. H. Hu, P. S. Hall, J. R. Kelly, P. Gardner, "UWB pyramidal monopole antenna with wide tunable band-notched behavior," Electronics Letters, vol. 46, no. 24, pp. 1588–1590, 2010.

[10] A. M. A. Salem, S. I. Shams, A. M. M. A. Allam, "A Miniaturized Ultra Wideband Antenna with Single Tunable Band-Notched Characteristics," In: Asia-Pacific Microwave Conference, Yokohama, pp. 746–749, 2010.

[11] S. A. Aghdam, "A novel UWB monopole antenna with tunable notched behavior using varactor diode," IEEE Antennas and Wireless Propagation Letters, vol. 13, pp. 1243–1246, 2014.

[12] S. A. Aghdam, J. Bagby, "Monopole antenna with tunable stop-band function," WAMICON, Orlando, FL, pp. 1–3, 2013.

[13] W. A. E. Ali, H. A. Mohamed, A. A. Ibrahim, M. Z. M. Hamdalla, "Gain improvement of tunable band-notched UWB antenna using metamaterial lens for high speed wireless communications," Microsystem Technologies, vol. 25, pp. 4111–4117, 2019.

[14] D. Sarkar, K. V. Srivastava, "SRR-loaded antipodal Vivaldi antenna for UWB applications with tunable notch function," International Symposium on Electromagnetic Theory, Hiroshima, pp. 466–469, 2013.

[15] S. Keerthipriya, C. Saha, J. Y. Siddiqui, Y. M. M. Antar, "Dual tunable multifunctional reconfigurable Vivaldi antenna for cognitive/multi-standard radio applications," IEEE International Symposium on Antennas and Propagation and USNC-URSI Radio Science Meeting, Atlanta, GA, USA, pp. 1833–1834, 2019.

[16] W. Jeong, S. Lee, W. Lim, H. Lim, J. Yu, "Tunable Band-Notched Ultra Wideband (UWB) Planar Monopole Antennas Using Varactor," In: European Microwave Conference, Amsterdam, pp. 266–268, 2008.

[17] T. Li, H. Zhai, L. Li, C. Liang, Y. Han, "Compact UWB antenna with tunable band-notched characteristic based on microstrip open-loop resonator," IEEE Antennas and Wireless Propagation Letters, vol. 11, pp. 1584–1587, 2012.

[18] M. Tang, H. Wang, "A novel ultra wideband planar antenna design with tunable band-notch characteristics," IEEE MTT-S International Microwave Workshop Series on Advanced Materials and Processes for RF and THz Applications (IMWS-AMP), Suzhou, pp. 1–3, 2015.

[19] Y. Li, X. Liu, K. Yu, Y. Wang, "A planar anti-interference UWB antenna with designated tunable and reconfigurable multiple filtering bands," Progress in Electromagnetic Research Symposium (PIERS), Shanghai, pp. 1977–1981, 2016.

[20] A. K. Horestani, Z. Shaterian, J. Naqui, F. Martin, C. Fumeaux, "Reconfigurable and tunable S-shaped split-ring resonators and application in band-notched UWB antennas," IEEE Transactions on Antennas and Propagation, vol. 64, no. 9, pp. 3766–3776, 2016.

[21] S. W. Wong, Z. Guo, K. Wang, Q. Chu, "A Compact Tunable Notched-Band Ultra-Wide-Band Antenna Using A Varactor Diode," Asia-Pacific Conference on Antennas and Propagation, Harbin, pp. 161–163, 2014.

[22] Z. L. Zhou, H. B. Zhang, L. Wang, D. G. Li, "A Novel Compact UWB Antenna with Dual Independent Tunable Band-Notches," International Conference on Microwave and Millimeter Wave Technology (ICMMT), pp.1–3, 2018.

[23] W. A. E. Ali, A. A. Ibrahim, "Tunable band-notched UWB antenna from WLAN to WiMAX with open loop resonators using lumped capacitors," ACES Journal, vol. 33, no. 6, pp. 603–609, 2018.

[24] M. Tang, H. Wang, T. Deng, R. W. Ziolkowski, "Compact planar ultra-wideband antennas with continuously tunable, independent band-notched filters," IEEE Transactions on Antennas and Propagation, vol. 64, no. 8, pp. 3292–3301, 2016.

[25] M. Nejatijahromi, M. Rahman, M. Naghshvarianjahromi, "Continuously tunable WiMAX band-notched UWB antenna with fixed WLAN notched band," Progress in Electromagnetics Research Letters, vol. 75, pp. 97–103, 2018.

# *Index*

Printed in the United States
by Baker & Taylor Publisher Services